be returned on or before
date below.

Advances in Additives for Water-based Coatings

**Cover illustration:** Microscopic view of a cross-section of a coated paper sheet. The sheet was coated with recovered coating colour obtained using the evaporative technique described in Chapter 8. The micrograph was taken using incident UV light. Hence, the areas of high brightness indicate areas of fluorescence and thus areas with a high concentration of optical brightening agent. This shows that the optical brightening agent present in the original coating colour has been successfully recovered using the evaporative recovery technique. (Micrograph by Jeremy Hooper ECC International; the micrograph copyright is owned by ECC International Ltd.)

# Advances in Additives for Water-based Coatings

Edited by

**G. Davison**
*Consultant, Chorley, UK*

**D.R. Skuse**
*ECC International, St. Austell, Cornwall, UK*

RS•C
ROYAL SOCIETY OF CHEMISTRY

The proceedings of a Symposium organized jointly by The Royal Society of Chemistry and the Society of Chemical Industry on 'Additives in Waterborne Coatings '97' held on 15 September 1997 at the University of York, UK

Special Publication No. 243

ISBN 0-85404-754-9

A catalogue record of this book is available from the British Library

© The Royal Society of Chemistry 1999

Published by The Royal Society of Chemistry,
Thomas Graham House, Science Park, Milton Road, Cambridge CB4 0WF, UK

For further information see our web site at www.rsc.org
Typeset by Computape (Pickering) Ltd, Pickering, North Yorkshire, UK
Printed by Athenaeum Press Ltd., Gateshead, Tyne and Wear, UK

# Preface

This review comprises selected papers from a symposium entitled 'Additives in Water-Based Coatings', which was held at the University of York in late 1997. The papers presented here are not intended to represent a comprehensive review of water-based additives but rather the intention is to highlight a significant advance in each of a number of important additive types.

The first chapter, by Hazel, discusses in depth the effect of volatile organic compounds [VOCs] on the ozone layer – something having a high current public profile – and the important economic implications for the development of low solvent emission coatings. This puts into perspective the current and proposed legislative environment, in its effect on the development of new coatings.

The following paper, by Walker, suggests that the move towards water-based systems requires, in general, more additives, in terms of the number of types and the volume required. Following a consideration of the influence of viscosity and of shear rate during use of different types of thickeners and the mechanism of their behaviour, the many factors affecting the choice of a thickener are examined.

Continuing, Doyle describes the use of recently developed synthetic silicates as rheology control agents and their use in a number of widely used waterborne systems, demonstrating their excellent compatibility with many commonly used polymer emulsions, pigments and extenders. The added benefit of antistatic properties on many substrates is noted.

The fourth chapter, from Schrickel, shows how the search for the optimum defoamer is, in essence, a compromise between compatibility and incompatibility in a particular system in order to achieve greater efficiency whilst taking into account a number of aspects which influence this choice. In this quest for optimisation a number of commonly used coating systems have been studied including wall, gloss and industrial (spray and heat cured) types as well as wood primers and printing inks. It is shown that the type of binder affects the behaviour of the defoamer in a particular formulation. It is concluded that the finding of the optimal defoamer is still an empirical process.

In Chapter 5 Randall reviews the role of coalescing solvents in the development of water-based formulations and describes the factors which enable the choice of a particular coalescing solvent to be made for use in different polymer systems. These highlight specific advantages of low volatility and odour together with rapid biodegradibility.

Waterborne formulations present, perhaps, potentially ideal conditions in many cases for the growth of microorganisms, and Weber shows, in Chapter 6, how contamination may arise, be detected and can then lead to the need for preservation in the coating. The chemistry of the many different chemical types of biocide available is considered, resulting in an 'efficacy spectrum' which assists in optimising the choice of the biocide required. The importance of good production hygiene and efficient microbiological control of raw materials is emphasised.

The penultimate chapter, by Hajas, introduces a new wetting and dispersing additive which is available in commercial quantities and whose performance is compared with conventional wetting and dispersing additives in emulsion polymers and in water soluble systems in general.

Finally, in Chapter 8, Skuse describes means by which direct raw material losses and those losses arising from waste colour during the processing of paper can be reduced using, for the latter, forced evaporation technology which offers many advantages over other colour recovery techniques. The recovered colours can be blended with fresh colour and used for their original duty with no significant decrease in performance. Clean water is a useful by-product of the process.

G. Davison
D.R. Skuse

# Contents

# Legislative Initiatives to Reduce Emissions of Volatile Organic Compounds

Nick Hazel

BP CHEMICALS, SALTEND, HULL, HU12 8DS, UNITED KINGDOM

## 1 Introduction

The coatings industry is in a state of change at the moment. European, national and regional Volatile Organic Compounds (VOC) emission legislation to limit ground level ozone is causing a *perceived* pressure to reduce the VOC content of coatings. However, there are two important things to bear in mind: firstly, it is the emission of solvents that is being regulated, which means that substitution options are not being mandated. Secondly, to claim environmental improvement, any VOC reduction should be considered on an 'as applied' or functional unit basis. Traditionally, consideration has only been given to formulation VOC content.

European legislation affecting the coatings industry actually forms only one part of a wider strategy to tackle the issue of atmospheric pollution. The wider perspective is important because the cost burdens on the coatings and other solvent using industries is being stacked up against the costs being borne by those affected by the other environmental programmes such as Auto Oil.

## 2 The VOC Issue in Perspective

First the good news: most solvents represent a negligible health risk once they get into the environment. The dilution levels are so high and the atmospheric lifetimes of the chemicals are so short that they never build up to cause health problems.

The next bit of good news is that the coatings industry as a whole is having no effect on the 'ozone hole' in the stratosphere. Ozone in the upper atmosphere provides a beneficial UV shield for all life on earth. Coatings solvents do not even reach the stratosphere, and spray can paints stopped using CFCs (the acknowledged culprits) some years ago.

Much closer to home, down in the troposphere (which is the air we breathe),

the same chemical, ozone, has attracted attention since it is a component of summertime smog

## 2.1 Ozone Sources

Figure 1 shows the major contributions to the ozone issue: clearly, ozone in the lower atmosphere comes from a variety of both natural and man-made sources.

There are two classifications of ozone at ground level, which relate to when and how it is formed.

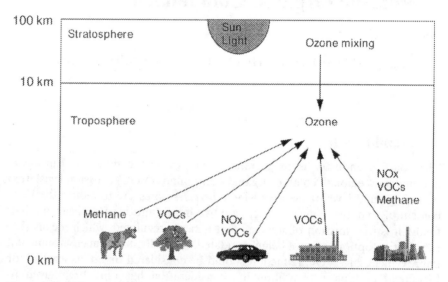

**Figure 1** *Contributors to the ozone issue*

*2.1.1 'Summertime Ozone'.* This is the issue driving current concerns over solvent emissions. Ozone peaks are produced during episodes of sunny weather. Technically speaking, the primary reaction to produce ozone is oxidation of oxygen by nitrogen dioxide, which in turn is reduced to nitrogen monoxide.

$$O_2 + NO_2 \rightleftharpoons O_3 + NO$$

VOCs are oxidized in the presence of sunlight and contribute by converting nitrogen monoxide into nitrogen dioxide. This disturbs the equilibrium by increasing the concentration of $NO_2$. VOCs can therefore increase the rate of ozone formation in sunny weather. The problem with summertime ozone is made worse because it adds to the existing background levels.

*2.1.2 Background Ozone.* Background ozone does not fluctuate strongly, and is not dependent on short-term changes in the weather, or seasons. The natural contribution to the background comes from ozone in the stratosphere making its way down to the ground, through a very slow mixing process. The man-made contribution comes from long-lived and relatively unreactive pollutants such as methane, CO and $SO_x$, in a similar way to VOCs, by oxidizing NO.

## 2.2 Ozone Precursors

Table 1 shows the distribution of atmospheric pollutants that contribute to ozone, from the CORINAIR-90 inventory. Since both VOCs and nitrogen oxides contribute to ozone episodes they are often grouped together under the term 'ozone precursor'. This is an important concept for comparing the EC Solvent Emissions Directive[1] (SED) with other environmental measures.

**Table 1** *Atmospheric pollution data for 1990: all ozone precursors (kte per year)*

| Europe (15) | $SO_2$ | $NO_x$ | VOC | $CH_4$ | CO |
|---|---|---|---|---|---|
| 1 Public power, *etc.* | 8 600 | 2 481 | 40 | 28 | 651 |
| 2 Combustion commercial, residential *etc.* | 1 270 | 499 | 730 | 369 | 6 775 |
| 3 Industrial combustion | 4 947 | 1 654 | 76 | 50 | 3 397 |
| 4 Production processes | 587 | 219 | 950 | 48 | 2 622 |
| 5 Extraction and distribution of fossil fuels | 44 | 76 | 1 039 | 4 846 | 61 |
| 6 Solvent use | | 1 | 4 085 | | 1 |
| 7 Road transport | 539 | 6 782 | 5 892 | 181 | 32 837 |
| 8 Other mobile sources and machinery | 420 | 1 678 | 536 | 20 | 1 924 |
| 9 Waste treatment and disposal | 77 | 117 | 240 | 7 329 | 2 755 |
| 10 Agriculture | 1 | 33 | 626 | 10 390 | 579 |
| 11 Nature | 573 | 49 | 3 547 | 9 056 | 1 327 |
| Total (for estimates listed) | 17 058 | 13 590 | 17 763 | 32 316 | 52 927 |

Source Eurostat/CORINAIR
Not all data is complete for all countries

## 2.3 Factors Influencing Ozone Concentration

Ozone is a highly reactive molecule; it can only persist in the atmosphere for a few days. Over time it will react with nitrogen monoxide and other oxidizable pollutants as well as natural materials it contacts. Ozone formation thus has a daily cycle; levels rise during the day as the photochemical production occurs, then fall at night as it reacts with other materials.

$NO_x$ is a primary pollutant that is being controlled in its own right. This is not simply for its role in ozone creation but also for its direct effects, as well as its role in acidification and to a lesser extent the role it plays in photochemical particulate formation (nitrates) and water eutrophication. Levels of $NO_x$ vary significantly across Europe; peak levels are associated with large urban centres.

The quality of ozone modelling has much improved in the last few years, partly due to better models and bigger computers, but also to more accurate emission inventories. Some of the findings are discussed below.

One very important point is the relative concentrations of $NO_x$ and VOCs, as one or the other may be the governing factor in ozone formation. For instance, in southern Europe, where there are high VOC levels (principally due to vegetation) and generally lower $NO_x$ (because of less intense industrialization and lower population densities), only changes in $NO_x$ have significant effects on ozone. This can be all the more significant when the proportion of man-made VOCs is examined, as well as their reactivity to give ozone.

Since most plant growth (which produces VOCs) occurs in the summer, natural emissions in the ozone season are grossly underestimated by Table 1. In fact plant growth is the strongest during exactly the same conditions that give rise to the peaks in ozone, and so occurs at the same time; their VOCs are also relatively reactive. Even with conservative estimates to correct this, one sees how important the natural contribution is (Figure 2).

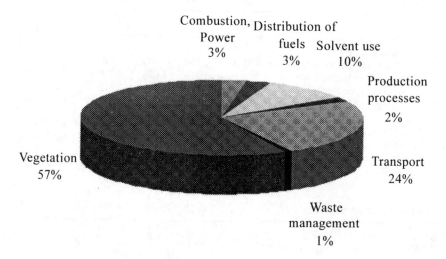

**Figure 2** *Summertime VOC emissions, reactivity weighted*

# 3 Legislative Measures and Agreements

Historically, the first international proposal on ground level ozone pollution was that embodied in the 1991 Geneva Protocol on Transboundary Pollution.[2] Although the EU and some member states were prompt signatories, it took until 1997 to be ratified. However, there are a number of important measures that focus on this problem.

## 3.1 Air Quality Standards

The Air Quality Management Framework Directive[3] lays the ground for a number of subordinate directives which set standards for air pollutants, *e.g.*

$NO_x$, $SO_x$ and ozone, but, importantly, not VOCs. Unlike for the other principal pollutants, the ozone proposal is expected to have only a target value, owing to the complexity of dealing with secondary pollutants; the rest will have binding limits.

The daughter directive on ozone (essentially the EU ozone strategy) is currently being drafted. Interested parties were involved in discussions with the EU Commission on the issue.

Air quality standards bring in the relatively new concept of non-attainment areas, and the prospect of a partially regional approach to reducing the ozone problem. However, they can also be a severe constraint on industrial growth in non-attainment areas.

## 3.2 The Integrated Pollution Prevention and Control Directive (IPPC)

This directive, now adopted, is one of the mechanisms the EU will use to achieve water quality standards, as well as air quality standards, for large installations. Essentially, it allows the regional authority to agree environmental priorities with larger installations to reduce the major problems, and avoid the transfer of emissions from one medium to another.

Industry has argued successfully for a harmonization of time-scales between IPPC and the Solvent Emissions Directive since some installations need to take action under both for the same emissions.

## 3.3 Auto Oil Programme (AOP)

A great deal of sophisticated atmospheric modelling[4] has been carried out to define the best options for the petrol and vehicle manufacturing industries to make their contribution to reducing the $NO_x$ and VOC levels (as well as some other atmospheric emissions). This modelling also predicts the expected ozone reductions from other legislation such as from control of power stations and solvent emissions.

Because most of us use cars, the AOP will have some direct effect on us personally, through the petrol we buy and engine/catalyst technology. As a result of this and other measures, urban air quality is expected to improve considerably over the next decade.

## 3.4 Solvent Emissions Directive (SED)

This is the proposed legislation that will be most familiar to coatings users and manufacturers. Some national legislation, such as TA Luft in Germany and the EPA in the UK, already pre-empts the aims of the SED. In the current proposals[1] there is flexibility for member states to build on existing national legislation to achieve the same ends (probably only for existing installations), rather than adopt the specific limits set out in it.

It seeks to control emissions from a wide variety of industrial sectors, not just coatings, and includes other solvent-using sectors such as rubber processing and seed oil extraction. However, coatings, ink and adhesive manufacturing and application industries do make up the bulk of the sectors in its scope.

So what has industry done and what is likely to happen in the future as a result of the SED?

# 4 Likely Effects of the Legislation

Before discussing how solvent-using industries are likely to adapt it is interesting to look at what has already been achieved over the last 15 years. Table 2 shows solvent inventory data for the UK for 1980 and 1995.

As can be judged from the data, these industries have considerably improved their emissions performance, during a period when there has been considerable growth in industrial output across Europe. Nearly all of this has been achieved for technical and economic reasons, and not through regulation. For this reason the steady improvement is expected to continue, as processes continue to become more efficient.

**Table 2** *UK inventory of solvent emissions* (kte/year)

| Solvent class | 1980 | 1995 |
|---|---|---|
| Propellants and blowing agents | 41 | 40 |
| Special boiling points | 61 | 30 |
| White spirits | 151 | 116 |
| Kerosenes | 40 | 31 |
| Isoparaffins & paraffins | 10 | 22 |
| Toluene | 33 | 30 |
| Solvent xylene | 78 | 65 |
| Other aromatics | 50 | 31 |
| Alcohols | 130 | 140 |
| Ketones | 84 | 90 |
| Esters | 31 | 71 |
| Glycol ethers, esters | 21 | 38 |
| Others | 3 | 2 |
| | | |
| Total non-chlorinated solvents | 733 | 706 |
| Chlorinated solvents | 149 | 73 |
| Grand total | 882 | 779 |

The data are also shown in Figure 3, which graphically illustrates which solvents are being preferred for the lower VOC coatings of today. Greater emphasis is being placed on coating performance, and less on price, these days.

So what of the future? The SED proposal provides two routes to compliance; these can be found in its Annexe III parts *a* and *b*. Essentially, the Annexe IIIa options require stack emission and fugitive limits to be met, and this is likely to be achieved largely through containment and end-of-pipe

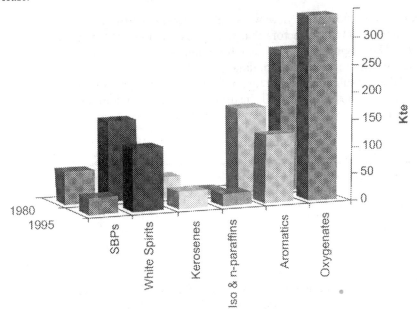

**Figure 3** *Changes in the pattern of use of solvents in the UK*

abatement. Annexe IIIb provides an alternative method for some industries, where a reduction scheme allows a flexible mix of options that include the above, and technology change (formulation and or application method).

Whether an enterprise will follow the Annexe IIIa or IIIb options will depend strongly on size, the type of operation and other individual circumstances such as performance requirements. What will happen is that the lowest cost option will be chosen *in each individual case.*

It is anticipated that not only are spends likely to be higher for small installations, but also Annexe IIIb options (substitution, *e.g.* by water-based formulations) will probably be more commonly selected at this scale. This reflects the fact that end-of-pipe abatement is more cost effective for larger installations.

## 5 Projected Costs of the SED

The European Solvents VOC Co-ordination Group (ESVOCC) has done considerable work to simplify and increase the flexibility of the SED. CEPE has played a crucial role in this achievement. As a result, the costs of the SED are about 40% lower than they would have been if the 1994 draft still stood.

Unfortunately, the earlier costing of the SED proposal[5] by the University of Karlsruhe missed some key points and the cost of the Directive is still high.

A more recent costing for the VOC Co-ordination Group has updated the

Karlsruhe study. This new work takes account of the considerable changes since the 1994 draft, sectors that were not costed and the increased size of the EU. The opportunity has also been taken to correct some of the earlier assumptions made for lack of data.

Table 3 summarizes the main findings from this cost study. However, current proposals from EU Institutions give some optimism that costs to industry will be lower than the figures in Table 3.

**Table 3** *Cost of the solvent emissions directive*

| | |
|---|---|
| Total emission reduction | 1 380 kte |
| Programme cash cost (20 yr) | 80 billion ECU |
| Programme cash cost (10 yr) | 55 billion ECU |
| Net present cost (NPC) | 50 billion ECU |
| NPC per te | 2 960 ECU |
| Marginal cost per te | 11 250 ECU |

Clearly, the true marginal cost of the SED is the maximum marginal cost borne by any of the industries in scope. This is probably in the region of 17000 ECU per te of avoided emission, or even higher, but the typical figure for SMEs of 11250 from Table 3 is probably a good guide.

To minimize its cost burden, each enterprise will have to investigate all the practical options open to it, exploiting the increased flexibility of the directive. The lowest cost option can then be chosen which allows existing quality requirements to be continued to be met.

These costs (both average and marginal) are broadly similar to those of the Auto Oil Programme, although a detailed comparison is difficult owing to the very different nature of the two industries.

Finally, what are the benefits that all of this will bring?

# 6 Benefits of the SED

The severity of ozone episodes across Western Europe has already declined over the last few years. This is due to industry improvements, including improvement of solvent use (Table 2) and regulatory measures adopted such as controls on $NO_x$ emissions from large combustion plants and the increasing proportion of cars with catalysts. The switch from coal to gas for power stations has also helped.

The effect of the new measures such as the SED and AOP can also be judged. Modelling work done for AOP and other studies indicate that over the next 10 years or so the number of days per year when the 90 ppb ozone information standard is exceeded will have fallen to an acceptable level. In short 90 ppb will not be exceeded 99% of the time in 99% of the land area of the EU, by 2010. In the meantime there is still a risk that many smaller enterprises, among them many coatings users, will have gone out of business by then.

# 7 Conclusions

The ozone issue is a very complex one and requires an integrated approach to controls, and much has already been achieved on all fronts through technical improvements.

The costs of the SED have been re-evaluated to take account of draft changes and the size of the EU. The current position is much more flexible and costs have been significantly reduced compared with earlier versions.

However, recent work shows the net present cost of the directive to be about 50 billion ECU: despite new optimism, there is still room for improvement, especially concerning high marginal cost measures.

# 8 References

1    Proposal for a Council Directive (EEC) on the Limitation of Emissions of Volatile Organic Compounds Due to the Use of Organic Solvents in Certain Activities and Installations, *Common Position EC No 40/98 16-June-1998.*

2    Geneva Protocol to the Stockholm Convention on Long Range Transboundary Air Pollution, *UNECE*, 1991.

3    Air Quality Management Framework Directive; *EC DGXI, OJ C216*, 6 August 1994

4    Auto Oil Programme, DGXI Press Release, 19 June 1996; *Air Health Strategy*, July 1996, p. 8.

5    O. Rentz *et al.*, Assessment of the cost involved in the Commissions draft proposal for a Directive on the limitation of the organic solvent emissions from the Industrial sectors, *DFIU*, Karlsruhe University.

# Thickeners for Water-based Systems: Current Technology and Future Advances

C. R. Walker

ALLIED COLLOIDS LIMITED, BRADFORD, WEST YORKSHIRE
BD12 0JZ, UK

## 1 The Thickener Market

The movement by coatings manufacturers to water-based systems has put new demands on thickeners and rheology modifiers.

The market for thickening agents in the UK is approximately 3500 dry tonnes per annum. The US market is approximately a factor of 10 larger and represents around 25% of the world consumption.

Although the coatings industry has recently shown growth of around 3% per annum and is likely to continue to do so for the next several years, the recent growth and predicted growth of thickening agents is closer to 5%. The reason for this is that the move towards water-based systems actually requires, on average, more additives both in terms of the number of additives and the volumes of each.

The market for solvent-free formulations is growing by approximately 9% per annum, fuelled by both increasing legislation and public demand, and shows no abatement.

## 2 Viscosity/Rheology

Viscosity modification is essential for almost every coatings formulation and in fact for many other formulations not related to the coatings industry, for example in the household products and personal care sectors.

Viscosity can be defined as the resistance to flow and varying degrees of resistance are required under varying degrees of shear if the formulation is to be deemed fit for purpose. Viscosity itself can be broken down into two components as per the following equation.

$$\text{Viscosity} = \frac{\text{Shear stress}}{\text{Shear rate}}$$

Shear stress can be thought of as the input of effort or energy, and shear rate as the flow output component. Hence if a lot of energy is put in but little or no flow occurs then the material could be considered to be 'thick'. The mapping of viscosity in relation to shear rate results in the rheology profile of the system.

Newtonian materials have a viscosity that is independent of shear rate, examples of which are mineral oils and pure solvents.

Dilatency is the increase in viscosity with increasing shear rate. It is a relatively rare phenomenon but one that is very useful, especially in the production of pigment grinds and slurries, where the high viscosities encountered at high shear rates aid the breakdown of aggregate and agglomerate particles.

Pseudoplasticity is the opposite of dilatency, *i.e.* decreasing viscosity with increasing shear rate, whilst thixotropy differs in that the rheological profile is shown to be time dependent and there is a recovery period before the material regains its original viscosity. Pseudoplastic and thixotropic rheologies are by far the most desired in formulations as will become apparent.

Every formulation has a theoretical optimum profile and the further away the actual profile is from the optimum the less 'fit for purpose' the formulation becomes. Dependent upon which areas of shear rate the current formulation is deficient in determines where the problems will be detected. For example, in Figure 1 if the dashed line is considered as the rheological optimum for this particular formulation then it can be seen that at low shear rates the

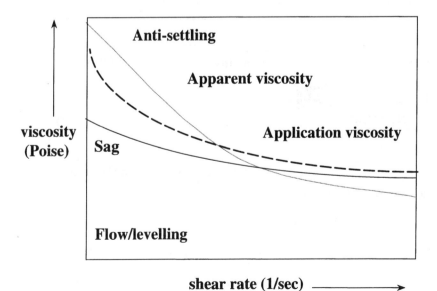

**Figure 1** *Rheological property* versus *shear rate*

formulation indicated by the solid line maintains a lower viscosity and hence variation from optimum manifests itself in the form of excessive flow after application. In addition, settlement problems may occur during the period before application.

In the same shear rate region the formulation indicated by the dotted line shows much higher viscosities and as such will not exhibit the problems associated with settlement or sagging. However, the flow characteristics after brushing, for example, will be inferior.

At medium shear rates, differences become less noticeable and it may well be that the consumer, upon shaking the containers, considers all three to be identical.

At the higher shear rates the dotted line again deviates markedly from the ideal and this low viscosity could, for example, result in the over-brushing of the product by an inexperienced consumer, leading to low coat weights and poor hiding power.[1-5]

## 3  Measurement of Viscosity

Various instruments are available to allow the formulator to measure viscosity. Some of the simpler instruments measure viscosity at a specific shear rate or across a narrow shear rate band. Examples are the Brookfield, Stormer, and Cone and Plate viscometers.

The Brookfield and Stormer viscometers measure viscosities that are indicative of the shaking and pouring of materials, best described as the apparent viscosity, that is to say apparent in the eyes of the consumer (Figure 2). The Cone and Plate viscometer, however, induces much higher shear rates and

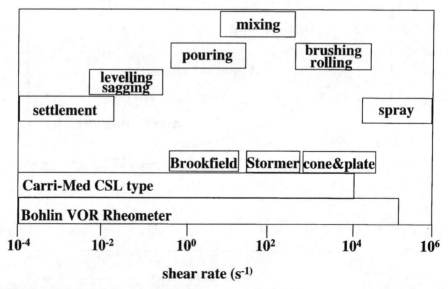

**Figure 2**  *Coatings rheology and viscometry*

simulates viscosities likely to be encountered during brushing, for example. All of the above instruments are commonly employed as quality control instruments.

Modern sophisticated viscometers measure across a much wider shear rate spectrum and allow graphical data to be produced. These instruments are of course much more expensive and are used predominantly as a research and development function.[6,7]

## 4 Thickener Types

Many chemical classes and subclasses of thickeners are commercially available. The main classes are:

1 Clays
2 Cellulose derivatives
3 Acrylics
4 Urethanes.

Within each class there are several subclasses (Figure 3).

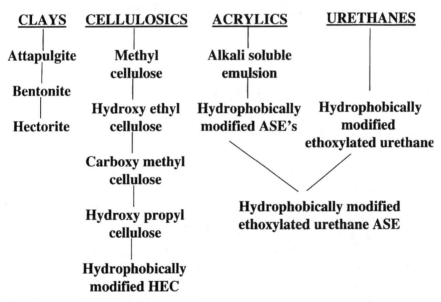

**Figure 3** *Thickeners – classes and subclasses*

Cellulosics include methyl cellulose, hydroxy ethyl and propyl cellulose as well as the anionic carboxymethyl cellulose and more recently hydrophobically modified HEC.

Acrylics were traditionally alkali soluble or swellable emulsions until, like the cellulosics, hydrophobic modifications were made.

Urethanes are the newest class known as hydrophobically modified ethoxylated urethanes and the most recent subclass of all is the HEURASE subclass – a blend of acrylic and urethane technologies.

## 4.1  ClayThickeners

These have already been comprehensively covered elsewhere in this volume and therefore will be mentioned only when comparisons are drawn.

## 4.2  Cellulose Derivatives

The non-ionic cellulose derivatives such as methyl cellulose and hydroxyethyl cellulose shown earlier are arrived at by the reaction of alkali cellulose with the relevant alkylene oxide and/or methyl chloride (Figure 4).

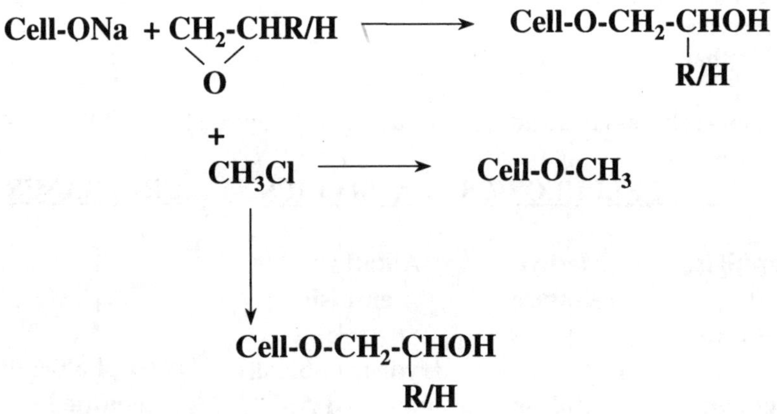

$$\text{Cell-ONa} + \underset{\underset{\text{O}}{\diagdown\diagup}}{\text{CH}_2\text{-CHR/H}} \longrightarrow \underset{\text{R/H}}{\text{Cell-O-CH}_2\text{-CHOH}}$$

$$+$$

$$\text{CH}_3\text{Cl} \longrightarrow \text{Cell-O-CH}_3$$

$$\downarrow$$

$$\underset{\text{R/H}}{\text{Cell-O-CH}_2\text{-CHOH}}$$

**Figure 4** *Cellulose derivatives*

The anionic sodium carboxymethyl cellulose is produced by utilizing sodium monochloroacetate as the reagent.

Differences between grades are essentially obtained by modifying the average number of substituted hydroxy groups and by the molar substitution of the anhydroglucose unit and also by the modification of the relative molecular mass ($M_r$). Blake[8] has highlighted how these modifications can influence properties such as spatter resistance, foaming tendencies, scrub resistance and colour acceptance.

These types of thickeners have had a strong market presence for the past three decades, offering a good blend of efficiency, rheology modification and gel structure that is highly desirable in latex paints and allied coatings. Hydrophobic modifications to cellulosics are now commercially available. They are produced by chemically bonding small amounts of aliphatic hydrocarbon moieties to the polymer backbone.

Traditional cellulose thickens because the hydrophilic polymer backbone forms hydrogen bonds with the water molecules and thickens the aqueous phase through entanglement of hydrated polymer chains. The moieties them-selves have hydrophobic tendencies and as such are capable of forming micelles in water. When chemically bonded to the hydrophilic backbone they retain their micellar tendencies.

The hydrophobic alkyl moieties form a micelle-like structure in the aqueous phase. These hydrophobic structures adsorb onto the surface of the latex binder particles when present in the system (Figure 5), leading to higher viscosities at high shear rates and allowing for a reduction in $M_r$ which then contributes to improved flow characteristics, *i.e.* a greater degree of New-tonianism is generated.

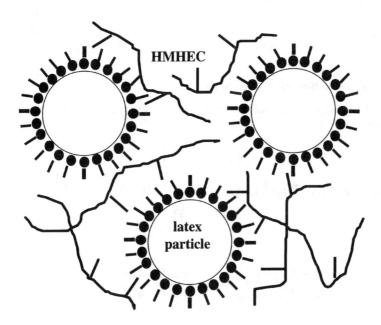

**Figure 5** *Thickening mechanisms of associative cellulose ethers*

## 4.3 Alkali Soluble Swellable Polymers

These products, also known simply as 'Acrylics', are also well established in the industry and, as with the cellulose based thickeners, are available with and without a hydrophobic modification for increased Newtonianism. A large number of design variables exist with these types of polymers and hence a spectrum of rheology profiles is possible (Table 1).

The introduction of associative modifying monomer in acrylics can have a large affect on the viscosity at the extremes of the normal shear rate spectrum while having no apparent effect on the in-can viscosity of the coating.

In addition, these types of products offer benefits in handleability as, being

**Table 1**  *Design variables for acrylic thickeners*

Degree of carboxylation
$M_r$
Polydispersity
Degree of covalent crosslinking
Degree of associative modification
Type of associative modification

low viscosity liquids, they are easy to pump and dose. They are not prone to bacterial attack and are very cost effective.

A paper presented in the US[9] has strongly suggested that although these types of products contain acid groups, they actually show much lower moisture affinities, measured by water vapour adsorption, than many people perceived.

## 4.4  Urethane Based Thickeners

These thickeners consist of hydrophilic and hydrophobic sections (usually terminal) (Figure 6).

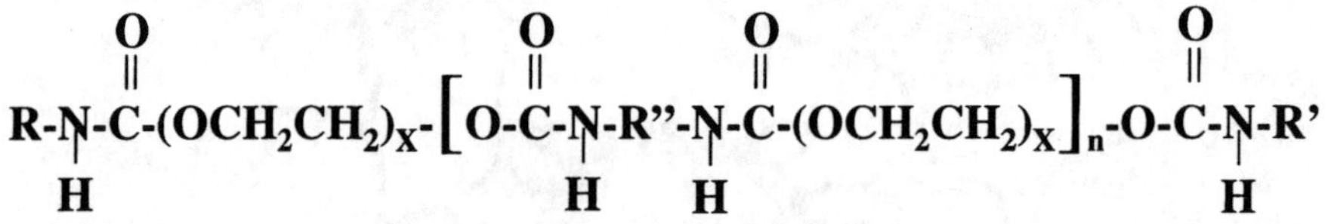

$$R-\underset{H}{N}-\overset{\overset{O}{\|}}{C}-(OCH_2CH_2)_X-\left[O-\overset{\overset{O}{\|}}{C}-\underset{H}{N}-R''-\underset{H}{N}-\overset{\overset{O}{\|}}{C}-(OCH_2CH_2)_X\right]_n-O-\overset{\overset{O}{\|}}{C}-\underset{H}{N}-R'$$

$$R = C_{12} - C_{18} \qquad X = 90 - 455$$
$$R' = C_{12} - C_{18} \qquad n = 1 - 4$$
$$R'' = C_7 - C_{36}$$

**Figure 6**  *Hydrophobically modified ethoxylated urethane*

The hydrophobic sections of these molecules associate strongly with various components of the formulation, especially the latex (Figure 7), and this high degree of association leads to these products having in general the least pseudoplasticity of all the classes discussed. This reduced pseudoplasticity exhibited manifests itself in coatings with excellent flow characteristics while imparting a much higher film build than many cellulosics or acrylics.

Formulation with these materials requires greater care since interactions with, especially, coalescing solvents and surfactants can have a significant impact on the final rheology of the coating.

Coalescing solvents can both increase and decrease the viscosity of a thickened latex and the viscosity change is primarily a function of the hydrophobicity of the solvent. Hence a hydrophilic material such as ethyl

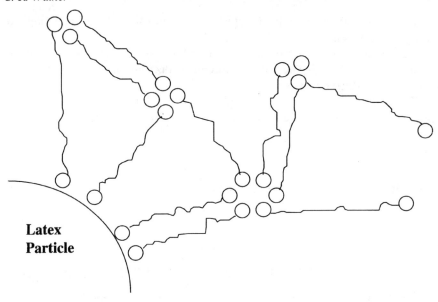

**Figure 7** *Hydrophobically modified ethoxylated urethane associative thickener network*

glycol causes a reduction in viscosity whereas Texanol (a non-water soluble alcohol) causes a viscosity increase.

Surfactants also modify viscosity at low and medium shear rates but to a lesser extent. The viscosity tends to decrease somewhat as HLB increases. Hence, anionic surfactants have the largest effect because of their hydrophilic nature.

The other potential drawback with these materials has been that they often contain solvent to suppress the viscosity but totally water-based products are becoming more and more commercial.

It is predicted that this class of thickener will grow since they are best suited to conferring the low degree of thixotropy synonymous with many current solvent-borne coatings.

## 4.5   Hydrophobically Modified Ethoxylated Urethane Alkali Soluble Emulsions

The HEURASE subclass of rheology modifier is the newest class of thickener on the market today. As their title suggests their composition and rheological properties are a compromise between acrylics and traditional polyurethane technologies.

This type of product is likely to find market acceptance as there was very little in the way of rheological overlap of the acrylics and urethanes before their introduction. Chemically they are terpolymers produced by emulsion polymerization of a carboxy functional monomer, a water insoluble monomer and a hydrophobically terminated urethane functional monomer. Product differentiation occurs mainly by altering the ratio of the monomers and, of course, $M_r$ adjustment.

The manufacturing technique is taken from acrylic technology and hence yields products that, like acrylics, require neutralization of the carboxy groups before solubilization will occur.

The thickening mechanism is still the classical hydrophilic/hydrophobic process but the hydrophobic 'associative' mechanism has been 'toned down' to lose some of the severe Newtonian contributions made by the original urethanes.

## 4.6 Other Factors Affecting Thickener Selection

The rheological properties conferred by the thickener to the formulation are obviously a key factor in product selection but it is far from being the only factor (Figure 8).

|                   | HEC | HMHEC | ASE | HASE | HEUR | HEURASE |
|-------------------|-----|-------|-----|------|------|---------|
| Thickening        | +   | +     | +-  | +    | +-   | +       |
| Roll spatter      | -   | +     | +   | +    | +    | +       |
| Brushing velocity | -   | +-    | +-  | +    | +-   | +       |
| Levelling         | -   | +-    | +-  | +-   | +    | +-      |
| Sag resistance    | +   | +     | -   | +    | +-   | +       |
| Water retention   | +   | +     | -   | +-   | -    | +-      |
| Colour acceptance | +   | +     | +   | -    | +    | +-      |
| Water resistance  | +   | +     | +-  | +    | +    | +       |
| Anti-blocking     | +   | +     | +-  | +    | +-   | +       |
| pH range          | +   | +     | -   | -    | +    | +-      |
| Cost              | +-  | +-    | +   | +    | -    | +-      |

**Figure 8** *Comparison of thickener systems*

Thickeners can also affect the formulation by altering:

(1) Cost
(2) Method of manufacture
(3) Water sensitivity
(4) Gloss
(5) Colour acceptance
(6) Spatter resistance
(7) Gel structure 'water retention'

Due to the huge number of rheological permutations possible as well as the many other factors that thickeners and rheological modifiers influence, the development of a truly universal thickener is extremely unlikely if not impossible.

The thickener and rheology modifier market has now developed to such an extent that rheology is now not an issue in its own right; that is to say that no matter what rheological profile is required for a particular formulation it can be attained by a product or combination of products already on the market with few exceptions.

It is important to remember that no one supplier company is rich in all classes of thickener. Even the major multinationals supply at maximum three classes, but two is more common, and as a result efforts are centred around covering as wide a rheological spectrum as possible with the classes available and on eliminating as many of the negative aspects as is possible.

To attempt to predict the future of thickeners we need to look at the coatings industry as a whole. The coatings industry is certainly not standing still. Global expansion is the priority for the larger companies looking to have access to a larger customer base. In fact over half of the top 10 Western paint companies made sizeable international acquisitions in 1996.[10]

Akzo Nobel purchased Nobiles Paint and Industrial Coatings business in Poland. Herberts took a controlling interest of Bitulac in France. ICI were particularly busy in the Americas. The US giants of Dupont, Sherwin Williams and PPG were also very active in South America and to a lesser extent Europe, mainly in the automotive and industrial coatings sectors.[11]

Global expansion brings with it rationalization and group purchasing.

These two things in turn have a negative effect on the number of suppliers and products on the market (including thickeners). Smaller manufacturers of thickeners can find that their valued customer is no longer interested until they manufacture and supply the same product in every major geographical area of the world and do it at a lower price!

For the future the market is likely to see new products predominantly of the polymeric classes where the technology is far from exhausted. Some of these products may well extend the rheological boundaries currently attained by existing products. However, it is far more likely that the products will be similar to those already on the market but will have been produced by smaller companies and will have been designed specifically for one or two medium-sized companies within a local market.

This is made possible because of the many new grades of latex binder entering the market *via* new manufacturing techniques and by the ever-increasing demand for higher performance water based systems as described earlier.

The fact that associative type thickeners are influenced to a much greater extent than their predecessors allows for minor modification of existing products to meet a customer's optimum requirement.

The business gained will be sufficient to satisfy the smaller producer's commercial requirements to enter the market, but not that of the majors to expand, and the business thus gained will aim to be sustained by out-performing the majors in areas of customer service and technical support. The latter is essential when formulating techniques with the various types of thickeners so diverse.

Whilst the introduction of new grades based on existing technology is relatively common place for all coating components, not just thickeners, the development of new subclasses and classes of products is of course much less frequent.

Allied Colloids longer term aims are to develop a new subclass of products that will marry together the positive points of acrylics and cellulosics whilst removing as many of the perceived negative points as possible in a similar fashion to those of the HEURASE.

In conclusion, the types of thickeners available today are diverse in terms of their chemistry, rheology and influence on other key properties of the final product. The use of two or even three thickeners in a system is the norm and, due to the quest for waterborne coatings of even higher performance, this is likely to continue with the emphasis on hydrophobically modified products to meet the rheological demands.

Product diversity will be tempered only by the rationalization and cost constraints imposed by the coatings manufacturer.

The quality of additional services such as technical support may well be the deciding factor as to whether a product survives in a market place that may be very active in the next several years.

# References

1   R.E. Smith, *J. Coatings Technol.*, 1982 (November), **54** (No. 694).
2   A. Heritage, *Paint Ink Int.*, 1990 (November), 22.
3   C. Rohn, *J. Water Borne Coatings*, 1987 (August), 9.
4   R.D. Athey, *Euro Coat.* 1991 (10), 672.
5   J. Haag, *Euro Coat*, 1992 (10), 641.
6   G. Schramm, *Euro Coat*, 1990 (10), 547.
7   P. Toepke, *Rheology 9l*, 1991 (June), 102.
8   D.M. Blake, *J. Coatings Technol.*, 1983 (June), Volume **55** (No.701).
9   G. Shay and K. Olesen, Stallings, *J. Coatings Technol.*, 1996 (March), **68** (No.854).
10  J. Bernie, *Coatings World*, 1996 (June), 14.
11  H. Jotischky, *Paint Polym. Coating J.*, 1997 (January), 24.

# Synthetic Silicates – Speciality Additives for Rheology Control of Waterborne Systems

Jane Doyle

LAPORTE INDUSTRIES, PO BOX 2, MOORFIELD ROAD, WIDNES, CHESHIRE WA8 0JU, UK

Laporte manufacture a range of synthetic silicates under the trade name Laponite which resemble the natural clay mineral hectorite in both structure and composition.

Laponite has two main functions:

- Rheology control agent
- Antistatic agent

Laponite has a number of distinct advantages compared with natural clays. Laponite is free from the silica impurities present in natural clay and is chemically pure and consistent in its properties as it is synthesized under carefully controlled conditions. Its efficiency as a rheology control agent is many times higher and it can be dispersed rapidly to give colourless, transparent, highly thixotropic gels without the need for high shear mixing, elevated temperatures or chemical dispersing agents.

Figure 1 shows the idealized structure of Laponite.

Laponite is a sodium magnesium lithium silicate with the following chemical formula:

$$Na^{0.7+}[(Si_8Mg_{5.5}Li_{0.3})O_{20}(OH)_4]^{0.7-}$$

This has a charge deficiency of 0.7 per unit cell. The idealized structure shown in Figure 1 would have a neutral charge with six magnesium ions in the octahedral layer, giving a positive charge of twelve. In practice, however, some magnesium ions are substituted by lithium ions and some spaces are empty to give this typical composition.

There are two types of Laponite available:

- Gel-forming grades and
- Sol-forming grades

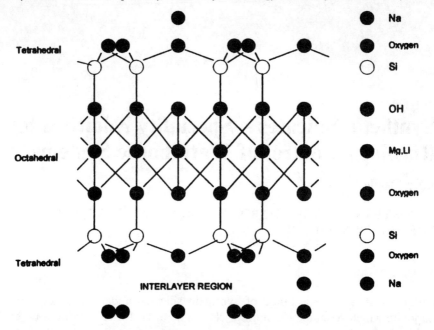

**Figure 1** *Idealized structural formula*

A gel is a high viscosity colloidal dispersion. A sol is a low viscosity colloidal dispersion. The processes occurring during the dispersion of Laponite into water are shown schematically in Figure 2.

A dilute dispersion of a gel-forming grade in deionized water may remain a low viscosity dispersion of non-interacting crystals for long periods of time. Electrostatic attractions draw the sodium ions in solution towards the crystal surface and osmotic pressure from the bulk of the water pulls them away. Equilibrium becomes established where the sodium ions are held in a diffuse region on both sides of the dispersed Laponite crystal. When two particles approach their mutual positive charges repel each other and the dispersion exhibits low viscosity.

The addition of polar compounds in solution (*e.g.* simple salts, surfactants, coalescing solvents *etc.*) will reduce the osmotic pressure holding the sodium ions away from the particle surface. This allows the weaker positive charge on the edge of the crystals to interact with the negative surfaces of adjacent crystals. This may continue, to give a 'house of cards' structure, which in a simple system of Laponite, water and salt is seen as a clear, colourless, thixotropic gel.

It is possible to modify Laponite from a gel-forming grade to a sol-forming grade by addition of certain compounds such as condensed phosphates, polyethylene and polypropylene glycols and certain non-ionic surfactants. In practice the sol-forming types of Laponite have been optimized by addition of tetrasodium pyrophosphate.

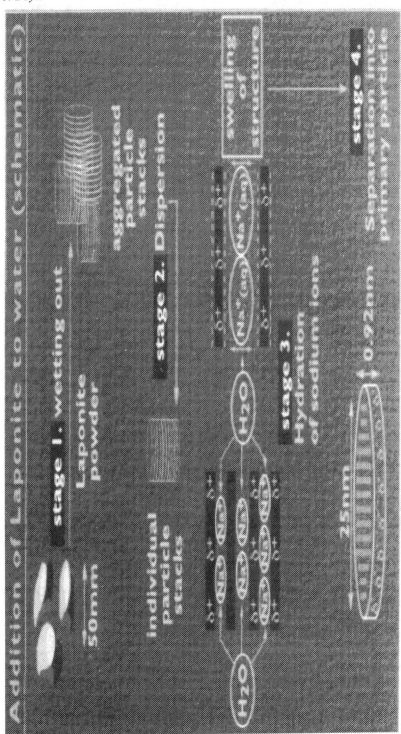

**Figure 2** *Dispersion of Laponite into water*

When a sol grade is added to water the Laponite will disperse as described earlier.

As the phosphate dissolves the pyrophosphate anions become associated with the positively charged edges of the Laponite crystal, making the whole particle negatively charged. This is subsequently surrounded by a loosely held layer of hydrated sodium ions, whose positive charges cause mutual repulsions between the dispersed Laponite crystals. This allows Laponite to be dispersed in water at higher concentrations, *e.g.* 12%, to give low viscosity sols.

The use of Laponite sols allows flexibility when formulating in that the sol may be incorporated into the formulation at any stage, *i.e.* the formation of structure may be delayed until a predetermined point during manufacture. They are also useful in hard water areas and where the water content of a formulation is very low. The main features and benefits of Laponite are detailed in Table 1.

**Table 1** *Features and benefits of Laponite*

| Feature | Benefit |
| --- | --- |
| Small primary particle size | Produces clear gels or sols to give ultra clear products<br>Disperses rapidly without the need for high shear |
| Synthetic layered silicate | High purity<br>Excellent consistency<br>Colourless dispersion |
| Inorganic | Cannot support microbial growth<br>Not affected by high temperature<br>Non yellowing<br>Non toxic<br>Non flammable |

With current trends and increasing environmental awareness leading to the replacement of solvent-based systems with waterborne systems, a wider formulation base for Laponite has developed. Laponite offers scope as a rheology control agent particularly in decorative, automotive, industrial, wood clearcoats and varnishes, and speciality coatings, including multicoloured paint and printing ink application areas. Laponite also offers key advantages in specialist applications such as conductive coatings.

In waterborne systems, rheology control agents are used to give the coating adequate storage stability and the correct rheology for application. Sufficient viscosity is required at low shear rates to prevent sedimentation of the pigments and to give the paint good in-can appearance.

However, the paint should ideally shear thin on application of shear, *i.e.* brush/spray to give easy application with sufficient film build and levelling before restructuring to prevent drips or sag. In practice it is often necessary to use combinations of thickeners to give the best results.

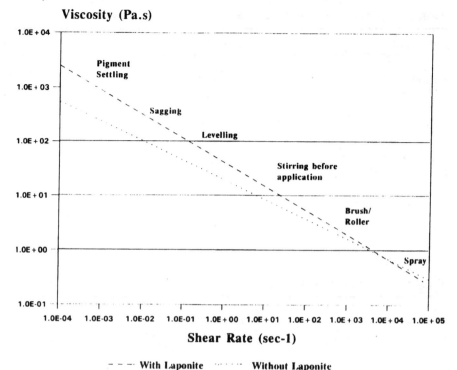

**Figure 3** *Effect of Laponite on paint viscosity*

The viscosity profile illustrated in Figure 3 shows how the addition of small quantities of Laponite (0.1%) can affect low and high shear rate viscosity.

The rheological properties of waterborne coating formulations depend on not only the thickeners used but also the complex interactions between these and the other components in the system, including pigments, extenders, binders, dispersants, surfactants and coalescing solvents.

The rust conversion coating formulation described in Table 2 demonstrates the use of Laponite RDS in conjunction with a co-thickener such as xanthan gum to give enhanced pH and viscosity stability.

A summary of the key features and benefits of Laponite as a rheology control agent in coating systems is shown in Table 3.

In the majority of coating systems, a key property that is required of the rheology control agent is the ability to suspend pigments. Laponite may be used to suspend a wide variety of pigments, including metallics, micas and inorganic pigments. The main advantages of using Laponite to stabilize pigments include:

- Low levels required (0.05–1.00%)
- Improved pigment wetting
- Stabilization of pigments without imparting thixotropy at low levels

**Table 2** *Rust conversion coating formulation*

| Raw materials | % Weight |
|---|---|
| Water | 11.09 |
| Laponite RDS | 0.75 |
| Defoamer | 0.20 |
| Xanthan gum | 0.20 |
| Tannic acid | 4.05 |
| Surfactant | 0.10 |
| Micronized barytes | 25.92 |
| Coalescent | 2.22 |
| Vinyl acrylic copolymer | 55.47 |
| Total | 100.00 |

| | |
|---|---|
| pH | 3.0. |
| PVC | 18.7%. |
| Cone & Plate | 0.8 Poise. |

**Table 3** *Application features and benefits*

| Feature | Benefit |
|---|---|
| High viscosity at low shear | Excellent suspension properties providing good appearance and syneresis control |
| Highly thixotropic | Readily formulated for application by brush, spray, dip, blade coaters and printing of inks<br>Good levelling and sag resistance properties<br>May be formulated to give enhanced film build |
| Interaction with other thickeners | Including cellulose ethers, natural clays and gums, polyurethanes and polyacrylic acid polymers<br>Synergistic effects including viscosity and stability improvements<br>Allows tailor-made rheological profiles |

- Does not effect gloss
- Laponite has been shown to reduce the release of hydrogen in aluminium based systems
- Long-term stability

The printing ink area is one area where suspension and stability of the pigment is paramount.

Printing inks can be divided into liquid inks and pastes. Liquid inks include flexographic and gravure. Pastes include screen and lithographic inks.

For example, carbon black is widely used in news ink. The main problem here is that the ink is diluted before printing, which causes the pigment to settle out. Laponite can be used here at low levels, which prevents settling of the pigment even on dilution.

Pearl lustre liquid inks based on mica pigments also exhibit stability problems. Using low levels of Laponite RDS can prevent settling without affecting gloss. This is demonstrated by the formulation described in Table 4.

**Table 4** *Pearl lustre printing ink formula*

| Raw materials | % Weight |
| --- | --- |
| Acronal 290D | 33.00 |
| Diofan 193D | 33.00 |
| Iriodin 300 gold pearl | 13.21 |
| Water | 15.49 |
| Dispersing agent N (10%) | 0.30 |
| Laponite RDS (10%) | 5.00 |
| Total | 100.00 |
| Ford cup No 4 = 35–40s. | |

In addition to its properties as a rheological additive, Laponite like many other colloidal materials is a natural film former. However, the unusual shape of the Laponite crystal, combined with its anionic nature, enables Laponite to produce films that have antistatic properties.

A film of Laponite can be cast onto paper simply from a dispersion of a Laponite sol grade in water. By selecting an appropriate binder and wetting agent system it is possible to produce coatings of Laponite on a wide range of substrates that are clear, highly flexible and moisture resistant.

Applications for Laponite as an antistatic agent include:

- Conductive layer in electrographic and antistatic paper
- High gloss, absorbent ink receiving coatings for ink jet printing
- Inert barriers in X-ray and photographic film
- Antistatic packaging for food and electrical components
- Fabric treatment
- Floor coverings – carpets

When coated onto a substrate, Laponite conducts electricity using two mechanisms:

- Electronic – the Laponite coating forms a continuous interlinked and overlapping film of electrically charged particles. This mechanism is not affected by changes in relative humidity.
- Ionic – a Laponite film will typically absorb up to 15% free moisture at 50% relative humidity. This is associated as water of hydration of the ions within the Laponite crystal structure. Some of this water is lost at low humidity. Laponite contains approximately 8% by mass of water, which is chemically absorbed into the crystal structure and may only be released at temperatures above 150 °C.

| $10^0$ | $10^5$ | $10^{12}$ | $10^{15}$ |
|---|---|---|---|
| conductive | static dissipative | | insulative |

**Figure 4** *Static dissipative range*

Figure 4 shows the range over which a coating is deemed to be conductive, static dissipative or insulative.

Laponite is compatible with a large number of water based resins, including acrylics, vinyl acetates, polyurethanes, melamines and epoxy esters and a wide range of permanent, clear, antistatic coatings have been developed.

Table 5 details a range of surface resistivities obtained of coatings prepared from different binders and applied to different substrate types.

**Table 5** *Antistatic applications of Laponite*

| Substrate | Laponite (%w/w) | Binder | Surface resistivity ($\Omega$ /square) |
|---|---|---|---|
| Paper | 10–20 | Styrene butadiene | $< 10^6$ |
| Transparency | 7.5–10 | Vinyl acetate ethylene copolymer | $< 10^6$ |
| Poly(vinyl chloride) | 7.5–10 | Polyurethane | $10^8$–$10^9$ |
| ABS/Styrene acrylic/ polycarbonate | 7.5–10 | Polyurethane | $10^8$–$10^9$ |
| Flexible films | 7.5–10 | Epoxy ester | $10^7$–$10^9$ |
| Glass | 7.5–10 | Melamines | $10^9$–$10^{12}$ |

In summary, Laponite is a versatile synthetic silicate used in a wide range of waterborne systems as a rheology control agent. Laponite products have been used in the coatings industry for over 30 years and show excellent compatibility with commonly used polymer emulsion systems, pigments and extenders. In addition to its rheological properties it is also an antistatic agent and forms highly conductive clear coatings on a wide range of substrates.

# References

1    *Laponite technical directory – Laporte Absorbents.*
2    J. Doyle and J. Barlas *PPCJ*, 1995(July), **185**(No. 4369), 15–17.
3    H. Van Olphen *An Introduction to Clay Colloid Chemistry*, 2nd edn, Wiley, New York, 1977.

# Defoaming Agents for Water-based Coatings: Influences of the Defoamer and Influences on the Defoamer

Jörg Schrickel

MÜNZING CHEMIE GMBH, HEILBRONN, GERMANY

## 1 Introduction

The introduction and stabilization of hydrophobic materials such as binder molecules, pigments and fillers into water-based coating systems has to occur through surface-active materials. Binder molecules for aqueous dispersions are stabilized by emulsifiers and pigments and fillers are incorporated by wetting and dispersing agents. All surface-active materials, however, tend to foam in aqueous systems.

Apart from the stabilization of bubbles by surfactant molecules, other factors exhibit an influence: formulation ingredients, production and application methods and, in addition, the substrate can support the creation of foam and increase or decrease the efficiency of a defoamer.

In a pure liquid, where no surfactants are present, air bubbles rise to the surface and burst. The interfacial tension between the air and the liquid is too high and the bubble cannot be stabilized. However, in systems where surface-active substances are present, air bubbles are stabilized like hydrophobic particles by surfactant molecules. These surfactant molecules form a layer around the bubble and lower the interfacial tension. The bubble still rises to the surface, but it does not burst due to the formation of a stable double layer with the surfactant layer on the surface. This layer consists of the monolayer of surfactant molecules at the air–liquid interface of the bubble and the monolayer of surfactant molecules that cover the liquid–air interface.[1–5]

According to the mechanism of foam formation the individual bubbles form a close spherical packing. Due to osmotic processes between the bubbles a drainage effect causes the removal of water from the interface between the bubbles. This water is collected in the void space between the bubbles. Due to this effect, a distortion of the spherical layer towards an octahedral geometry can occur by forming a narrower distance between the double layers. This so-called polyhedral foam consists of hexagonal close packed air.[6,7]

Two other effects allow the foam to stabilize. The Marangoni effect describes the tendency for a process in the case where one part of the surface lacks surfactant molecules. This lack causes an increased surface tension and the system attempts to restore the former surface tension by migration of surfactant molecules.

The second effect is an electrostatic repulsion of surfactant molecules from two different monolayers. Due to the narrowing of the two layers in the lamellae, through drainage effects, the two layers may collapse and burst the bubble. This narrowing is prevented by the electrostatic forces of charged parts of the surfactant molecules, which keep the monolayers at an equilibrium distance.

In practice, the described stabilization of a foam to give a stable polyhedral foam only occurs on very few occasions in paints and similar coatings during production and application. Normally, problems arise during production and application due to individual bubbles that lead to surface defects in the applied film. Modern defoamer technology helps to overcome foam problems because the mechanism of foam formation and destruction is understood. The number of possible ingredients, and therefore the high number of their combinations, have until now prevented the development of a method to predict the required type of defoamer needed for a given formulation or problem. Additionally, influences from the system on the defoamer increase the uncertainty for a defoamer prediction. Empirical work, however, has led to great experience in the application of defoamers and helps in finding the right defoamer.

# 2 Foam Formation

## 2.1 Formation of Foam during Production

During the stirring and dispersing process air is introduced into coating systems. Vigorous stirring leads to the incorporation of air and its stabilization due to surfactants. Air can also be incorporated by pigments and fillers due to their rough surface structure. At this stage of production it is important to use an efficient dispersing and wetting agent to remove this captured air from the pigment surface and the defoamer must be efficient for the transport of air outside the system.

The requirements for a defoamer during production depend on the characteristics of the coating system and the type of production. During the grinding process, where high shear forces are applied, vigorous stirring is required and a large amount of air is incorporated. A highly efficient, and therefore hard to incorporate, defoamer has to be used. A defoamer that does not resist high shear forces would be split into small droplets, which are inefficient in this system. Whereas in low viscous, low pigmented systems the use of hard-to-emulsify defoamers might cause problems in the applied film surface and might tend to separate, in this case easy-to-emulsify defoamers have to be used.

Apart from these influences, from an economical point of view, production times are longer if a foam is formed during production. A large amount of trapped air in the grinding process will reduce viscosity and therefore shearing becomes less efficient.

In the letdown phase, where temperature and shear sensitive binders are added, the stirring is less vigorous. Here, the addition of defoamers prevents the formation of foam and removes foam that was stabilized by the emulsifier of the binder.

## 2.2 Formation of Foam during Pumping and Filling

During processing (pumping, filling, transport) air can also be introduced into the system due to movement at the air/liquid interface. The coating is also pumped, transported and moved in different application methods such as printing, curtain coating or dipping. Not only the foam that is formed during production but also that produced later on, while pumping and filling, has to be destroyed. The defoamer therefore also has to be efficient during processing.

To meet quality control requirements a constant density of the coating is necessary for a constant filling of the cans. Furthermore, the optical aspect is important as customers do not appreciate paint cans with foam on the surface. The defoamer therefore has to be active during storage.

## 2.3 Formation of Foam during Application

The methods of application and the kind of substrate have a strong influence on the tendency for foam formation during application. The application tool (brush, roller, spray gun, *etc.*) has an influence that is just as important as the substrate. Especially porous substrates, such as walls, wallpaper, untreated wood, *etc.*, show an increasing tendency to form foam.

Spray application leads to high requirements for the performance of a defoamer since the coating is saturated with air after leaving the nozzle. The defoamer must be efficient in immediately destroying any microfoam formed on the substrate.

# 3 Defoamer Formulation

In general, defoamers consist of several compounds and these compounds are not compatible with the coatings medium. Problems arise with defoamers that are too compatible in the system, which means that they are inefficient. Otherwise, if they are too incompatible in the system, these defoamers might be highly efficient but they will cause problems in the applied film. The ideal solution in the application of defoamers is to find the best balance between compatibility and incompatibility of the defoamer for maximum efficiency.

Defoamers are multicomponent systems with different groups of functional ingredients. There are three principal classes of components: carriers, which make up 75–90% of the system, hydrophobic ingredients, which make up 5–10%, and other special substances (0–20% of the formulation).

Carriers are different kinds of oils in general; the most versatile and least costly are mineral oils. Beside these, paraffinic (medical white) oils are used to meet health regulations (BGA/FDA) and for food and potable water applications. Furthermore, vegetable oils (biologically degradable and for low VOC paints), silicone oils, polysiloxanes and water are used as transporters. In minor amounts, fatty acids, fatty alcohols and poly(vinyl ethers) are used.

The purpose of the carrier is to spread on the surface to remove the layer of surfactant molecules. Therefore, it has a lower surface tension than water. Its other important purpose is to carry the hydrophobic ingredients from the defoamer formulation to the double layer. Therefore, they have to be insoluble and incompatible with the solvent (water) to rise to the surface. This monolayer formed by the defoamer ingredients must have a reduced surface elasticity compared with the original surface elasticity of the foam bubble double layer.

Hydrophobic ingredients can be used in a liquid or solid form. In a liquid form they exist as emulsion droplets. Particles are always hydrophobic materials with a particle size between 0.1 and 20 μm. The particle size is an important factor in terms of entry into the double layer. If the particle size is too small, the effectiveness of the particle is reduced significantly because water can enter into the double layer and dilute the concentration of the defoamer particles. If the particle size is too big, the particle can not enter into the double layer and a collapse of microbubbles into macrobubbles can not occur.[8–10]

The most important task of hydrophobic ingredients is to absorb surfactant molecules from the double layer, by which the surface tension is increased and the bubble bursts. The compounds used as hydrophobic particles are waxes, hydrophobic silica, metal soaps and also polypropylene glycols, amides and polyurethanes.

The concentration of the particles in a defoamer formulation is limited due to the fact that the total concentration of surfactant molecules has to be kept over a particular limit in order not to destabilize the whole emulsion, which would lead to a flocculation of the polymers.

Emulsifiers are of great importance because they regulate the compatibility of the defoamer in a coating system. Depending on the emulsifier, the defoamer is dispersed to a certain extent and develops an efficiency that depends on the particle size. Therefore, the emulsifier is the most important ingredient in finding a balance between the compatibility and incompatibility of the defoamer in a coating system. Furthermore, as emulsifiers are also surface-active materials it is desired that they have a low tendency to foam.

Other materials present in a defoamer can be biocides, thickeners or protective colloids.

# 4 Experimental

## 4.1 Experimental Test Methods

*Red Devil Test.* The 'red devil test' is the standard test method to measure the air incorporated in coatings of both low and high viscosity systems. Samples are shaken with different types of defoamers in a red devil mixer, which allows intensive shaking in a reproducible way during which air is incorporated into the sample. Defoamer concentrations and periods of shaking can be adapted to necessary requirements. The density of the sample, as a function of incorporated air, is determined after shaking and compared with the value of an untreated sample without incorporated air. The values are given as percentage of incorporated air. The test can be repeated with already shaken samples that have been stored to determine long-term stability of the defoamers.

Standard: 70–100 g sample/ 0.2–0.4% defoamer/ 5 min (up to 50 min) shaking time.

*Dissolver.* For all kinds of coatings, but especially for systems of lower viscosity, the dissolver test is performed. A 50 ml sample is placed in graduated glass cylinders of 250 ml. The sample is stirred with a toothed dissolver disc for a certain period of time. The foam formation is measured just after stirring and the resistant foam is determined after a certain period of time. The first value indicates the efficiency of a defoamer to prevent the formation of foam whereas the second value indicates the efficiency in destroying the foam already formed. After storing these samples for a certain period it is possible to judge the tendency of separation of the defoamer.

Standard: 250 ml cylinder/50 ml sample/0.2% defoamer/2800 rpm/ 3 min stirring time/ quantity of foam is determined immediately after stopping the dissolver and 60 s later.

*Levelling Glass Plate.* Coating samples obtained from the red devil or dissolver test are applied with defined film thickness on a glass plate with a blade (25–100 μm). Evaluation of the surface aspect is done under wet and dry conditions to judge the compatibility between the defoamer and the coating system (pinholes, orange peel, craters, fisheyes and other defects).

*Roller and Brush Application.* A defined amount of paint is applied on PVC foil, PE foil, wood, paper *etc.*, depending on the purpose of the coating. The application tool is a highly structured foam roller or a brush. The creation of foam during application is judged as well as the efficiency of destruction both during and after drying.

*Spray Application: with airless/air-mix spray gun.* Spray application is performed for industrial lacquers and all other spray applied coatings. During the application there is a high saturation of air in the coating. The resulting surface properties are evaluated to determine the compatibility and the ability to destroy microfoam.

*Gloss.* Certain defoamer components can exhibit gloss-reducing properties. Especially, hydrophobic silica and waxes in high concentrations may reduce the gloss. Gloss values are measured in gloss paints to avoid the recommendation of a defoamer that reduces gloss.

*Surface Analysis.* A quantitative method used to evaluate the quality of an applied film is the analysis of film defects. Film defects in the dry film are determined on a defined surface. A small number of defects indicate a good compatibility of the defoamer.

# 5 Results and Discussion

A large number of different kinds of defoamers are available on the market and it is difficult for paint and adhesives producers to find the right choice of defoamer for a particular application. Suppliers have always tried to facilitate the selection process and to support the customer. Apart from the defoamer, all other additives and ingredients exist in many forms and are supplied by many producers. For every formulation therefore a high number of possible variations of ingredients results, and changes of products in a guide formulation can change the behaviour of the final product significantly. Due to mutual influences of different kinds of products, their consequences have to be considered and, therefore, empirical knowledge is needed.

As demonstrated in Table 1 different types of defoamers are tested in various formulation types. The performance of the defoamers in these formulations is completely different. Type 1 defoamer is very efficient (++) in a wall paint formulation and gives a good performance (+) in the wood primer formulation but in the other cases its behaviour is only average (0). The best defoamer performance in the gloss paint formulation is given by defoamer type 2 and the recommendations for the other formulations are given accordingly. A symmetrical pattern of recommendations is generated.

**Table 1** *Defoamer type* versus *application I*

|  | type 1 | type 2 | type 3 | type 4 | type 5 |
|---|---|---|---|---|---|
| Wall paint | ++ | + | + | 0 | ++ |
| Gloss paint | 0 | ++ | 0 | 0 | + |
| Industrial paint | 0 | + | ++ | + | + |
| Wood primer | + | − | − | ++ | + |
| Printing ink | 0 | + | + | + | ++ |

Performance: ++ very good; + good, 0 fair; − low.
type 1: silicone free mineral oil defoamer, hard to incorporate, long term stability, stable to separation.
type 2: silicone free, pharmaceutical grade oil defoamer, easy to incorporate.
type 3: polysiloxane emulsion defoamer, easy to incorporate.
type 4: silicone oil compound, easy to medium incorporation.
type 5: polysiloxane defoamer, hard to incorporate, high efficiency, and high performance.

However, general information cannot be gathered from Table 1 because this pattern was generated on purpose. In Table 2 an order is introduced, listing the defoamers in order of increasing defoamer technology and the application in order of decreasing polymer content (solid). No symmetrical pattern can be obtained in this case and predictions seem to be much more difficult.

**Table 2** *Defoamer type* versus *application II*

|  | type 1 | type 4 | type 2 | type 3 | type 5 |
|---|---|---|---|---|---|
| Industrial paint | 0 | + | + | ++ | + |
| Gloss paint | 0 | 0 | ++ | 0 | + |
| Printing ink | 0 | + | + | + | ++ |
| Wall paint | ++ | 0 | + | + | ++ |
| Wood primer | + | ++ | − | − | + |

Performance: ++ very good; + good, 0 fair; − low.
type 1: silicone free mineral oil defoamer, hard to incorporate, long term stability, stable to separation.
type 2: silicone free, pharmaceutical grade oil defoamer, easy to incorporate.
type 3: polysiloxane emulsion defoamer, easy to incorporate.
type 4: silicone oil compound, easy to medium incorporation.
type 5: polysiloxane defoamer, hard to incorporate, high efficiency, and high performance.

On the following pages the influences on and of the defoamer will be discussed in examples, leading to a guideline for solving defoamer problems.

## 5.1 Wall Paint

The formulation characteristics of wall paint (Table 3) are medium to high PVC values, a low polymer content and a long open time. Apart from the $TiO_2$ pigments a large amount of fillers are used in wall paints. The PVC can be greater than the CPVC. In general, cellulosic ethers are used as thickeners to keep the open time long enough.

**Table 3** *Formulation*

| | | |
|---|---|---|
| *Binder* | Vinyl acetate/veova/acrylic acid ester | 35.0 |
| | $NH_3$ | |
| *Defoamer* | Type 1 | 0.4 |
| *Water* | | 1.9 |
| *Dispersing agent* | Polyacrylic acid | 0.3 |
| *Co-dispersant* | Polyphosphate | 1.5 |
| *Thickener* | Cellulosic ether, grade 6000, (3% solution) | 6.0 |
| | HMHEC (3% solution) | 5.0 |
| *Pigment* | $TiO_2$ | 13.5 |
| *Fillers* | Calcium carbonate | |
| | Clay | |
| | Calcite | |
| | Talcum | |
| | Al-silicate | Total 34.0 |
| *Biocide* | | 0.2 |
| *Solvent* | Mineral white spirit | |
| | Butyldiglycol acetate | 2.1 |
| *Total* | | 100.0 |
| PVC: | 49%. | |
| Polymer content (solid): | 17.5%. | |

During production, transport and filling of such a paint the volume and weight of the product have to be kept constant. To avoid flowing over during production, weighing differences and foam (in the can and during application), an efficient defoamer has to be added during grinding. A low content of foam also helps to keep dispersion times short.

During application no foam should be formed and when it is it should be destroyed immediately. The customer's behaviour also has to be taken into account as wall paints are often diluted with 10–20% water before application. To improve the scrub resistance the formation of microfoam should be suppressed. For low emission wall paints special defoamers are available.

As can be seen from Table 4 the destruction of air after 5 min, 25 min and after storing with repeated red devil tests can be done without big differences with all kinds of the tested defoamers. A better performance may be given by these types of defoamers, which are difficult to incorporate into a system. This means that they resist the applied shear forces during the grinding process. As a result, especially in application, they still have enough performance to destroy the foam and the bubbles that are produced by roller application.

**Table 4**  *Wall paint: test results*

| *Tests:* | Red devil | 5 min |
|----------|-----------|-------|
|          |           | 25 min |
|          | Storage   | Samples from red devil 5 min, 14 days at 40 °C, Then 25 min red devil test |
|          | Levelling glass plate | Compatibility |
|          | Roller application | Foam during application |

|         | *Air content (%)* *5 min* | *Air content (%)* *25 min* | *Storage* *25 min* | *Levelling* | *Roller* *application* |
|---------|---------------------------|----------------------------|--------------------|-------------|------------------------|
| type 1  | 1.1 | 4.2 | 5.7 | Good     | Very good |
| type 2  | 1.5 | 4.4 | 5.9 | Average  | Average   |
| type 3  | 1.3 | 6.5 | 6.8 | Good     | Average   |
| type 4  | 1.5 | 5.4 | 5.3 | Very bad | Good      |
| type 5  | 1.1 | 4.2 | 5.9 | Bad      | Very good |

Wetting agents, dispersants, thickeners and the pigmentation show a bigger influence on the defoamer in such coating systems than the binder does. Therefore, it has to be taken into account that the exchange of a formulation component can change the performance of a defoamer. As the paint industry wishes to formulate several paints with the same defoamer, a universally applicable defoamer is required with good overall properties.

## 5.2 Gloss Paint

Gloss paints (Table 5) show an increasing percentage of polymer content and a decreasing percentage of pigments and fillers compared with wall paints. Different types of pigments (organic, inorganic, specially coated, *etc.*) are used and therefore rising quantities and high-performance wetting and dispersion agents are required in gloss paint formulations. Parallel to a rising polymer content and a high content of wetting and dispersing agents the influences of these formulation components on the defoamer increase.

**Table 5** *Formulation*

| | | |
|---|---|---|
| *Solvent* | Propylene glycol | 4.9 |
| *Wetting agent* | Maleinic acid copolymer | 0.8 |
| *NH₃* | | 0.1 |
| *Biocide* | | 0.1 |
| *Pigment* | TiO$_2$ | 19.6 |
| *Defoamer* | Type 2 | 0.2 |
| *Binder* | Acrylate | 56.8 |
| *Solvent* | Methoxy butanol | 2.5 |
| Coalescent | Ester alcohol | 3.6 |
| *Thickener* | Acrylate/water 1:1 | 5.5 |
| *Water* | | 5.9 |
| *Total* | | 100.0 |
| PVC: | 17%. | |
| Polymer content (solid): | 24%. | |

Gloss paints are applied for decorative and protective reasons. They are applied by brush or roller but can also be spray applied. Therefore, the specific requirements for defoamers are higher than for wall paints.

Defoamers used for gloss paint applications (Table 6) should be easily incorporated without leaving surface defects. To obtain a better aspect of the applied film, gloss paint formulations often contain a silicon oil defoamer. Silicon oils exhibit better spreading properties on the surface due to lower surface tension values compared with mineral oils.

The differences in the air content values are bigger than in the wall paint formulation, showing the bigger influence of the binder and wetting agent. There are also significant influences from the defoamer on the gloss. Defoamer type 2, which gives the best gloss, additionally shows the best surface properties obtained by application of the red devil samples, which were shaken 5 and 25 min, respectively.

## 5.3 Industrial Paint (Spray Applied, Heat Curing Metal Lacquer)

Industrial paints (Table 7) are similar to gloss paints but more demanding application methods have led to a higher performing defoamer formulation. For special industrial application methods (such as dipping, curtain coating, spraying, *etc.*) and drying procedures (UV curing, heat curing, 2K lacquers, *etc.*) special binder types have been developed. Therefore the influences of binders, wetting and dispersing agents as well as other formulation components on the defoamer are high.

**Table 6** *Gloss paint: test results*

*Tests:*  Red devil            5 min  
                                  25 min  
        Levelling glass plate    Compatibility  
        Gloss 60 °            From glass plate application  
        Roller application     PVC foil, foam during application  
        Surface analysis       [Number of surface defects on 25 mm$^2$]  
                                  samples derived from shaking 5 and 25 min in the red devil mixer

| | Air content (%) 5 min | Air content (%) 25 min | Levelling | Gloss 60° | Roller application | Surface analysis 5 min | Surface analysis 25 min |
|---|---|---|---|---|---|---|---|
| type 1 | 1.9 | 3.3 | Very good | 70.7 | Good | 18 | 44 |
| type 2 | 1.7 | 2.0 | Very good | 81.4 | Good | 15 | 35 |
| type 3 | 2.1 | 5.8 | Very good | 76.0 | Very good | 24 | 61 |
| type 4 | 3.4 | 8.7 | Very good | 78.8 | Good | 23 | 36 |
| type 5 | 1.4 | 6.6 | Very good | 80.0 | Very good | 15 | 48 |

**Table 7** *Formulation*

| | | |
|---|---|---|
| *Binder 1* | Acrylate 1 | 35.0 |
| *Solvent* | Butyl glycol | 0.4 |
| *Pigment* | TiO$_2$ | 16.6 |
| *Defoamer* | Type 3 | 0.2 |
| *Dispersing agent* | Polymeric dispersant | 0.8 |
| *Neutralizing agent* | NH$_3$ | 0.2 |
| *Binder 2* | Acrylate 2 | 35.0 |
| *Binder 3* | HMMM | 6.3 |
| *Solvent* | Butyl glycol | 2.5 |
| *Water* | 3.0 | |
| | | |
| *Total* | | 100.0 |
| PVC: | 10%. | |
| Polymer content (solid): | 38%. | |

The dry film of an industrial lacquer has to give a perfect surface due to the protective and decorative purpose of industrial applied paints and it also has to satisfy the customer's demands. High-performing defoamers (Table 8) like type 3 have to be used to meet these requirements. Therefore it is important not to allow foam formation during application or at least to destroy the foam immediately and completely where it is formed. A long-term efficiency and stability is desired to prevent a loss of efficiency on the application machine. A sufficient wetting of the substrate is also important due to the fact that metal and plastic surfaces are often found as substrates in industrial applications.

The defoamer for an industrial coating must be a high-performance defoamer, for example a polysiloxane emulsion type in order to obtain an easy

**Table 8** *Industrial heat curing metal gloss paint: test results*

| *Tests:* | Red devil 5 min | Air content in % |
| | Levelling glass plate, 100 µm | Compatibility |
| | Gloss 60° (blade) | Glass plate |
| | Roller application | Foam formation during application |
| | Levelling spray | Metal plate |
| | Gloss 60° (spray application) | Metal plate |
| | Surface analysis | [Defects/25 mm$^2$], metal plate |

| | *Air content (%) 5 min* | *Levelling glass plate* | *Gloss 60° Blade* | *Roller application* | *Levelling spray appl.* | *Gloss 60° spray appl.* | *Surface analysis* |
|---|---|---|---|---|---|---|---|
| type 1 | 2.5 | Average | 89.8 | Bad | Good | 90.5 | 31 |
| type 2 | 4.6 | Average | 93.6 | Average | Very good | 93.8 | 25 |
| type 3 | 2.7 | Very good | 94.7 | Good | Very good | 95.9 | 23 |
| type 4 | 2.5 | Average | 93.8 | Good | Bad (pinholes) | 94.2 | 65 |
| type 5 | 4.1 | Good | 94.6 | Bad | Very good | 94.9 | 26 |

incorporation without formation of substrate defects such as craters and pinholes. The defoamer has to show long-term stability without giving problems of incorporation at post addition.

Even though the defoamer of choice does not give the best absolute defoaming result the overall properties are the best. Defoamer type 3 shows a very good compatibility on the glass plate application and the best gloss values after blade application on the glass plate and after spray application on a metal plate. The surface analysis of the spray applied samples shows that microfoam, which often appears after spray application, is efficiently reduced.

## 5.4 Wood Primer

Primers and impregnations can vary in viscosity, pigment and filler content and polymer content, depending on their end use and application. A general treatment is therefore not possible and in this case the example of a low viscous, unpigmented wood impregnation was chosen.

The formulation displayed in Table 9 contains a high content of binder, coalescent and solvents, which means that together with surfactants a high formation of foam can be expected.

The defoamer of choice (Table 10) has to be easy to incorporate but efficient during the production and application of the primer. During storage the tendency to separate has to be suppressed to obtain a homogeneous product.

As impregnations are overpainted it has to be assured that no problems will arise while overpainting the first primer layer. In the special case of wood as a substrate an impregnation has to give protection from UV, water and humidity; it has to protect from outside as well as from inside and additionally contains biocides, fungicides, *etc.*

Due to the low viscosity the defoamer test is run with a dissolver in a graduated glass cylinder to measure the volume of the produced foam after

**Table 9** *Formulation*

| | | |
|---|---|---|
| *Binder* | Styrene acrylate | 40.0 |
| *Wetting agent* | Aminomethyl propanol | 0.2 |
| *Solvent* | Propylene glycol | 3.0 |
| *Coalescent* | Ester alcohol | 0.6 |
| *Biocide* | | 2.0 |
| *Wetting agent* | Polysiloxane polyether | 0.5 |
| *Defoamer* | Type 4 | 0.2 |
| *Coalescent* | Ester alcohol | 0.2 |
| *Water* | | 53.3 |
| *Total* | | 100.0 |
| PVC: | 0. | |
| Polymer content (solid): | 14%. | |

**Table 10** *Wood primer: test results*

| *Tests*: | Dissolver test | 0.3% Defoamer |
|---|---|---|
| | | 0.5% Defoamer |
| | Levelling glass plate | Compatibility |
| | Wood application | Foam formation on rough surface |
| | Defoamer separation | (Dissolver samples; 24 h) |

| | *Dissolver test foam (ml) 0.3%* | *Dissolver test foam (ml) 0.5%* | *Levelling* | *Separation* | *Wood application* |
|---|---|---|---|---|---|
| type 1 | 10/0 | 10/0 | Good | Medium | Very good |
| type 2 | 100/90 | 70/60 | Average | Strong | Good |
| type 3 | 210/180 | 180/120 | Average | Low | Good |
| type 4 | 90/10 | 30/10 | Good | Very low | Very good |
| type 5 | 150/130 | 10/5 | Average | Very low | Very good |

stirring. The first value indicates the foam immediately after stirring and the second value gives the value 1 min after having stopped the dissolver. The performance of the defoamers to prevent the formation of foam (first value) and to destroy the formed foam (second value) is not the most important point. To prevent a separation of the defoamer in the can during storage it has to show a good compatibility. The best foam killer (type 1) exhibits a medium tendency to separate whereas defoamer types 4 and 5 show a very low separation tendency.

## 5.5 Printing Inks

Printing ink formulations (Table 11) contain in general a high content of binder and solvent, but also a great variety of different pigment types are used. The content of wetting and dispersing agents therefore is high and viscosities can vary significantly, depending on the kind of printing ink.

**Table 11** *Formulation*

| | | |
|---|---|---|
| *Binder* | Acrylate | 55.6 |
| *Pigment* | TiO$_2$ | 27.7 |
| *Wax emulsion* | PE emulsion | 8.8 |
| *Water* | | 2.7 |
| *Solvent* | Isopropanol | 5.0 |
| *Defoamer* | Type 5 | 0.2 |
| *Total* | | 100.0 |
| PVC: | 23%. | |
| Polymer content (solid): | 23%. | |

The most demanding task for a defoamer in a printing ink is its application (Table 12). Due to high application rates, short distances, short transition times on various transport media and a continuous pumping process the creation of foam can easily take place.

**Table 12** *Printing inks: test results*

| *Tests:* | Red devil | 5 min |
|---|---|---|
| | | 25 min |
| | Levelling glass plate | Compatibility |
| | Levelling PE foil | Compatibility, wetting properties |
| | Roller application | Simulation printing process |

| | *Air content (%)* *5 min* | *Air content (%)* *25 min* | *Levelling Glass plate* | *Levelling PE* | *Roller application* |
|---|---|---|---|---|---|
| type 1 | 2.9 | 5.8 | Average | Average | Average |
| type 2 | 2.1 | 4.6 | Very good | Good | Good |
| type 3 | 1.8 | 4.9 | Good | Good | Average |
| type 4 | 2.0 | 3.3 | Very good | Very good | Good |
| type 5 | 1.4 | 3.1 | Very good | Very good | Very good |

The requirements for an efficient defoamer for printing ink formulations are resistance to the applied shear forces, easy incorporation, immediate destruction of foam on all transport media and on the substrate. It also has to be appropriate for post-addition.

The most efficient defoamer for printing applications is a pure polysiloxane type defoamer. It is highly efficient in destroying the foam and shows a very good compatibility. In the application simulated by roller application it also exhibits the best results. As it is difficult to incorporate it exhibits a long-term efficiency.

# 6 Influences on the Defoamer

Apart from different formulation types and their peculiarities in application other influences can determine the efficiency of a defoamer. In this chapter the effects of isolated properties such as binder type, emulsifier system or dispersing agent on the defoamer performance are studied.

## 6.1 Influence of Different Binder Types

The binder type can have a strong influence on the performance of a defoamer. If only the efficiency of a defoamer in a dispersion is tested and further effects such as separation or stability are ignored, strong differences in the performance will be obtained.

Table 13 shows the results of experiments in which several different defoamers are tested in different binder types. Some 70 g of dispersion was treated with 0.2% of defoamer and shaken for 5 min in a red devil shaker. The given values are % air content.

**Table 13** *Influence of different binder types*

|  | Acrylate | Acrylate-Copolymer | Vinyl acetate Veova-Acrylate Terpolymer | Alkyd resin emulsion | Vinyl acetate emulsion | Acrylate-Urethane |
|---|---|---|---|---|---|---|
| AGITAN® 230 | 15.4 | 14.8$^a$ | 11.5 | 29.3 | 11.6$^a$ | 12.9$^b$ |
| AGITAN® 260 | 19.9$^a$ | 8.2$^b$ | 15.4$^a$ | 29.4$^a$ | 11.3 | 19.1 |
| AGITAN® 295 | 2.0$^b$ | 8.7 | 6.5$^b$ | 21.5 | 4.2$^b$ | 19.6$^a$ |
| AGITAN® 760 | 5.8 | 9.2 | 6.9 | 13.2$^b$ | 6.4 | 16.5 |

Values: air content (%). Test conditions: 70 g emulsion, 0.2 % defoamer, 5 min red devil.

$^a$ Worst. $^b$ Best.

As can be seen from the table the performance of the defoamers in different binder types can vary widely. In the pure acrylate the efficiency of the best defoamer (AGITAN® 295) is 10 times better than that of the worst (AGITAN® 260). In the acrylate–urethane binder, differences in the performance of the defoamers are smaller and in this special case AGITAN® 295 shows the worst result.

As can be seen from this investigation no general prediction can be made that indicates the best defoamer type for a special binder type.

## 6.2 Influence of Different Emulsifier Systems

The influence of the kind of binder on the performance of a defoamer is already strong but even no general recommendations can be given for one type of binder that is produced by different suppliers. In the case of the pure acrylic binders in Table 14 the emulsifier system and the content of emulsifier in the binder has a tremendous influence on the defoamer performance.

In acrylate 1 the AGITAN® 295 shows the best result with low air content levels. The difference between this and the worst result, AGITAN® 260, is

**Table 14** *Influence of different emulsifier systems*

|  | Acrylate 1 | Acrylate 2 | Acrylate 3 | Acrylate 4 | Acrylate 5 | Acrylate 6 | Acrylate 7 |
|---|---|---|---|---|---|---|---|
| AGITAN® 230 | 4.5 | 45.47 | 3.5 | 5.4[a] | 10.8 | 5.1[a] | 9.7[a] |
| AGITAN® 260 | 4.6[a] | 39.0[b] | 4.8[a] | 5.3 | 7.3 | 3.7 | 7.2 |
| AGITAN® 295 | 3.7[b] | 51.3[a] | 0.8[b] | 0.5[b] | 12.1[a] | 1.4[b] | 5.2 |
| AGITAN® 760 | 3.9 | 47.4 | 1.9 | 1.2 | 2.8[b] | 3.3 | 3.9[b] |

Values: air content [%]. Test conditions: 70 g emulsion, 0.2% defoamer, 5 min red devil.

[a] Worst. [b] Best.

low. In acrylate 2, the values are 10 times higher and the performance of AGITAN® 260 is the best. The subsequent examples (acrylates 3–7) exhibit the fact that no general predictions can be made.

## 6.3 Influence of the Formulation

A strong difference in the performance of defoamers is observed on comparing the results obtained from defoamer tests in a pure dispersion with those in a formulated paint containing this binder. Table 15 shows the results of the defoamer tests of this comparison.

**Table 15** *Influence of the formulation*

|  | Defoamer Air content (%) | Dispersion Levelling | Lacquer Air content (%) | Roller application |
|---|---|---|---|---|
| AGITAN® 230 | 11.3 | Average | 2.6 | Average |
| AGITAN® 256 | 5.1 | Average | 5.1 | Good |
| AGITAN® 260 | 9.8 | Good | 1.8[a] | Good |
| AGITAN® 295 | 5.2 | Average | 4.4 | Very bad |
| AGITAN® 655 | 15.3[b] | Average | 3.1 | Very bad |
| AGITAN® 700 | 9.1 | Average | 7.1[b] | Bad |
| AGITAN® 760 | 3.9[a] | Good | 2.8 | Good |

Dispersion: mixture of acrylates  values: air content (%)
Lacquer: window metal lacquer

[a] Best [b] Worst

For dispersion, AGITAN® 760 shows the best defoaming properties of the defoamers displayed in the table. If the lacquer is produced out of this dispersion and the defoamer test repeated with these defoamers the best result is obtained with AGITAN® 260. This defoamer had shown only average results in the pure dispersion. In the lacquer, AGITAN® 260 is 5 times more efficient whereas AGITAN® 760 only shows an improvement of 1.4 times. AGITAN® 230 also shows an interesting behaviour, changing from the second worst to the second best defoamer in the lacquer, with an improvement that is greater than a factor of 4.

The formulation components such as wetting and dispersing agents, pigmentation, type of thickener, *etc.* have a strong influence on the performance of a defoamer and no general derivation can be made, as demonstrated in Table 15. Only an experimental test can give the required information.

## 6.4 Influence of the Dispersing Agent

Wetting and dispersing agents play a crucial role in the foaming properties of a paint formulation as they tend to foam or create foam due to their chemical nature. On the other hand they have an important influence on the defoaming properties in pigmented systems.

One substantial property of wetting and dispersing agents is the wetting and coverage of the pigment surface with the wetting and dispersing agent, respectively. This results in the liberation of captured air which is adsorbed on the pigment surface. The more efficient the wetting of the pigment is the more air will be liberated during the dispersion process. In combination with an efficient defoamer at this stage of paint production an important content of air can be removed from the system.

Table 16 shows three high-gloss paints, containing $TiO_2$ as pigment, that have been prepared with different dispersing and wetting agents but without defoamers. The properties of the paints are evaluated by visual inspection and by determination of the density. The absolute density is given as well as the relative density based on the completely defoamed air-free system.

**Table 16** *Influence of the dispersing agent*

| Defoamer | polyacrylate | polyacrylate + non-ionic wetting agent | polymeric dispersant |
|---|---|---|---|
| *Visual aspect after dispersion* | Very foamy | Foamy | Foamless |
| *Density* | $d = 1.06$ g ml$^{-1}$ | $d = 1.14$ g ml$^{-1}$ | $d = 1.25$ g ml$^{-1}$ |
| *Relative density* | $d* = 0.81$ | $d* = 0.902$ | $d* = 0.99$ |
| | *Air content (%)* | *Air content (%)* | *Air content (%)* |
| AGITAN® 230 | 4.4 | 4.1 | 0.6 |
| AGITAN® E 255 | 18.2 | 13.8 | 7.6 |
| AGITAN® 650 | 5.9 | 4.7 | 3.1 |
| AGITAN® 760 | 4.8 | 4.7 | 0.4 |

Formulation type:    high gloss paint; carboxylated acrylic binder.
PVC:    15%.
Polymer content:    32%.

A polyacrylate dispersing agent leads to a very foamy paint after preparation, which is confirmed by the measurement of the density and comparison with the completely defoamed system.

If a wetting agent is also added the foam after preparation is reduced, which is confirmed by the determined densities. The high-performance polymeric dispersing agent leads to a foamless paint after dispersion in which 99% of the final foam-free density is obtained.

The defoamer tests show that the efficiency of the defoamers rises with the performance of the dispersing agent. The defoamers in the paint prepared with the polyacrylate dispersant show only moderate efficiency. The efficiency of the best defoamer is 4 times better than the worst, indicating low specification. Similar results are found for the combination of classical dispersant and wetting agent, with the best defoamer having a 3 times higher efficiency than the worst.

In the paint prepared with the polymeric dispersing agent, much lower air content values are found and the difference between the best defoamer and the one with the lowest efficiency is a factor of 19. As displayed from this result the efficiency of the defoamers in combination with a high-performance dispersing agent can be increased. As a side effect in this experiment (normally it is the most important result obtained from the dispersant) the polymeric dispersing agent gives much higher gloss values than the polyacrylate dispersing agent.

The values of air content are related to the corresponding series of dispersing agent and cannot be compared as absolute values between these series. In accordance to the test method air is incorporated into the system during the defoamer test.

# 7 Conclusions

The search for the right defoamer for a coating system is a question of balance between compatibility and incompatibility. A defoamer in general is an incompatible ingredient in a coating system. The efficiency is related to the incompatibility and the more efficient a defoamer is, the less compatible it is. Highly efficient defoamers are just compatible enough not to cause surface defects or other problems in the coating. A defoamer that is too compatible loses its efficiency. Therefore, the best compromise between efficiency and compatibility has to be found.

It has been demonstrated in this paper that a great number of factors determine the efficiency of a defoamer. Most of the formulation ingredients play an important role for the efficiency of the defoamer but the application method and the substrate also influence the formation of foam. The defoamer itself exhibits influences on both the properties and the aspect of a coating.

Due to the complexity of all these mutual interactions it is still impossible to predict the appropriate defoamer by theoretical assumptions. The finding of an optimal defoamer remains an empirical process[11,12] in which a large number of determining factors should be taken into account. Apart from the list given in Table 17, in which a large number of different binders were tested in different applications and for which the best defoamers were determined, Münzing Chemie GmbH offers the possibility to conduct customer's defoamer tests in the laboratory.

**Table 17** *Defoamer recommendation: architectural paints/coatings and plasters*

| Product name | Supplier | Application | Defoamer recommendation | Additive recommendation |
|---|---|---|---|---|
| ACRILEM 30 WA<br>Vinylversatate/Vinylacetate copolymer | Icap, Italy | Paints, plasters | 0.2% AGITAN 260 or 295 or 701 | TAFIGEL PUR 40 |
| ACRILEM ST 190<br>Styrene-acrylate copolymer | Icap, Italy | Wall coatings, wood and metal paints | AGITAN 260 or 701 | TAFIGEL PUR 40 or PUR 60 |
| ACRONAL 18 D<br>Acrylic/Methacrylic acid ester Cop | BASF | House paints, renderings, wood coatings | 0.5% AGITAN 295 or 281 | |
| ACRONAL 290 D<br>Acrylic acid ester/Styrene Cop. | BASF | Paints, plasters, filler | 0.4% AGITAN 280 or 260 or 0.2–0.4% AGITAN 315 or 731 | |
| ACRONAL 296 D<br>Acrylic acid ester/Styrene Cop. | BASF | Paints, plasters, filler | AGITAN 281, 315 or E 256 | |
| ACRONAL 355 D<br>Acrylate | BASF | Sealing compounds, construction adhesives | AGITAN 232 or 295 or 296, or 315 | |
| ACRONAL DS 6210<br>Acrylic acid ester/Styrene Cop. | BASF | Elastic coatings | AGITAN 731 or 760 or 700 | |
| ACRONAL S 610 (DS 6149)<br>Butylacrylate/Styrene Cop. | BASF | Plasters, indoor paints, silicate paints and plasters | AGITAN 731 or 315 | |
| ACRONAL S 716<br>Styrene-Acrylate Cop. | BASF | Facade paints, indoor paints, plasters | AGITAN 760 or 315 | |
| Alberdingk SC 44<br>Styrene/Acrylic acid ester Cop. | Alberdingk Boley | Indoor/outdoor coatings for mineral substrates | 0.4% AGITAN 260 or 0.2% 295 | |
| Alberdingk SC 46<br>Styrene/Acrylic acid ester Cop. | Alberdingk Boley | Indoor/outdoor coatings for mineral substrates | 0.4% AGITAN 260 or 280 | |
| Alberdingk SC 420<br>Styrene/Acrylic acid ester Cop. | Alberdingk Boley | Indoor coatings | 0.4% AGITAN 260 or 0.2% 295 | |
| Alberdingk SC 4400<br>Styrene/Acrylic acid ester Cop. | Alberdingk Boley | Indoor/outdoor coatings, flat and gloss systems | 0.2% AGITAN 701 or E 256 | |
| CRILAT D 117<br>Acrylate cop. anionic | VINAVIL, Italy | Wall coatings | AGITAN 230, 281, 731 | |
| CRILAT D 120<br>Styrene ester/acrylate cop. | VINAVIL, Italy | Wall coatings | AGITAN 217, 230, 260, 701, 731 | |

# Acknowledgement

I would gratefully like to thank our co-workers in the customer service laboratory for their support and especially the head of the laboratory, Mr. Peter Bissinger.

# References

1　W. Gress, *Swiss Chem.*, 1992, **14**, 59.
2　W. Heilen, O. Klocker and J. Adams, *PPCJ*, 1996, **1**.
3　D. Stoye (Ed.), *Paints, Coatings and Solvents*, VCH, Weinheim, 1993.
4　L.J. Calbo (Ed.), *Handbook of Coatings Additives*, Marcel Dekker, New York, 1987, Vol. 1.
5　D.R. Karsa (Ed.), *Additives for Waterbased Coatings*, The Royal Society of Chemistry, Cambridge, 1991.
6　J.A. Kitschner and C. F. Cooper, *Quart. Rev.*, 1959, **13**.
7　J.A. Kitschner, *Recent Progress in Surface Science*, Academic Press, New York, London, 1964, Vol. 1.
8　A. Prins, *Food, Emulsions Foams*, 1987, **58**, 30.
9　K. Koczo, L. A. Lobo and D.T. Wasan, *J. Colloid Interface Sci.*, 1994, **492**, 150.
10　E. Wallhorn, W. Heilen and S. Silber, *Farbe und Lack*, 1996, **12**, 30.
11　W. Schultze (Ed.), *Dispersions-Silikatsysteme*, Expert Verlag, Renningen-Malmsheim, 1995.
12　W. Schultze (Ed.), *Wässrige Siliconharz-Beschichtungssysteme für Fassaden*, Expert Verlag, Renningen-Malmsheim, 1997.

# Recent Advances in Coalescing Solvents for Waterborne Coatings

David Randall

CHEMOXY INTERNATIONAL PLC, MIDDLESBROUGH TS3 6AF, UK

## 1 Introduction

This review is intended to cover the development of water based coatings, the polymer systems used, the role of coalescing solvents in these formulations, the factors which affect the choice of coalescing solvents and finally to indicate the recent developments within my own company of esters of low volatility, low odour and rapid biodegradability for use in this area.

The first task of any reviewer of this field is to define the terms he will be using. The term paint is widely used to describe a coating applied to a variety of substrates for protective or decorative purpose. 'Paint' also implies a pigmented species, whereas 'surface coatings' is a broader term for coating systems with or without pigments used for any coating purpose. Since the majority of the systems I shall be reviewing will normally be pigmented, I hope that my somewhat indiscriminate use of both terms will be forgiven.

All paint systems may be considered as a combination of a small number of constituents. These are

Continuous phase:      Vehicle (polymer and diluent)
Discontinuous phase:   Pigment (pigment and extender)
                       Additives

This article is not concerned with the details of paint formulation, but the polymer types most widely encountered are Alkyds, Polyurethanes and Nitrocellulosics in solvent based systems, and Acrylics, Styrene–Acrylics and copolymers of vinyl acetate in water based coatings.

## 2 Water Based Coatings

Until recently, most paints were solvent based. Since many polymers were produced in solution it was natural that the coatings system derived from these polymers would use the solvent of reaction as the diluent for the polymer. These were almost invariably organic species. The objections to the use of organic solvents in paint formulations were at first confined to those with strong views regarding the loss of significant quantities of organic species to the environment. Latterly, these opinions have been reinforced by the role of some organic species in damage to the ozone layer, or to the production of smogs in the lower atmosphere. This is as a result of the reaction between organic species present, which is catalysed by sunlight and exacerbated by the presence of other pollutants associated with motor vehicles *etc.* Flammability of paints also poses a significant problem in storage and use.

All this tended to reinforce the need to develop aqueous based systems, for which these problems would be eliminated.

## 3  Emulsion Polymers

Of course, the manufacture of polymers need not be carried out in solution; emulsion polymerization has a long and honourable tradition in the field of macromolecular chemistry. This polymerization technique is performed using water-insoluble monomers, which are caused to form an emulsion with the aqueous phase by the addition of a surfactant. Polymerization may be effected by the use of a variety of radical-generating initiating species, and the relative molecular mass ($M_r$) of the polymer is controlled by the use of chain transfer agents and the concentration of initiator employed. The final polymer dispersion is often described as a latex, named after natural rubber, which is also an emulsion polymer!

This polymerization technique allows for the formation of copolymers in which the addition of relatively small quantities of comonomer may have a significant effect on the final properties of the polymer. This is particularly the case with the glass transition temperature ($T_g$) of the polymer. This is a physical transition that occurs in the polymer when the amorphous structure of the polymer begins to change from a glassy to a rubbery state. At temperatures below a polymer's $T_g$, it will be relatively brittle, and will be unlikely to form a coherent film.

This effect has a major impact on the use of polymers in aqueous systems. In the case of a polymer in solution, the presence of the solvent plasticizes the polymer during film formation. A polymer with a high $T_g$, *i.e.* one greater than ambient temperature, can, when in solution, be applied at temperatures below its $T_g$. In the case of an emulsion polymer, water is a non-solvent in the system and film formation below the polymer $T_g$ is unlikely. The temperature at which a coherent film may be formed from a solution or emulsion based system is known as the minimum film forming temperature (MFFT or MFT).

Figures 1–4 describe the formation of a discrete film from an emulsion system.

When the film is applied to the substrate, discrete polymer particles are dispersed in the aqueous phase. The system is stabilized by the presence of

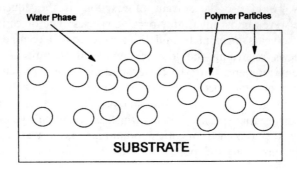

**Figure 1** *Latex in contact with substrate*

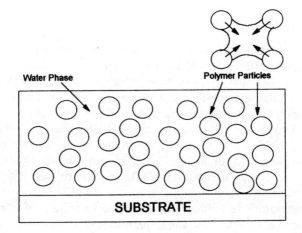

**Figure 2** *Initial water evaporation and film formation*

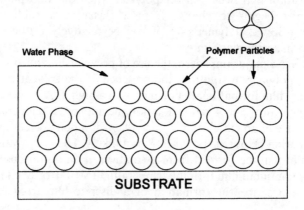

**Figure 3** *Close packing of latex particles*

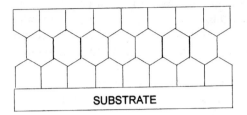

**Figure 4** *Coalescence to form a polymeric film*

the surfactant at the water-particle interface. The particles are spherical, with an average diameter of 0.1–0.2 μm. As water is lost from the system, the mutual repulsive forces associated with the surfactants present inhibit the close packing of the particles and a cubic arrangement of the particles is formed.

As the water continues to evaporate, the particles become close packed with a solids volume of around 70%. Capillary forces continue to force the particles together.

The final stage is achieved when most of the water is lost from the system. Here, the interparticular repulsive forces are overcome by increasing surface tension and the particles coalesce into a discrete film. This will only occur at temperatures in excess of the MFFT.

# 4 Role of a Coalescing Solvent

Most coatings based on emulsion polymers are used in environments where they will be expected to form a coherent film at temperatures as low as 0 °C. However, other physical properties besides film-forming capability are required from the polymer. These include abrasion resistance, hardness, chemical resistance, impact performance, *etc.* These can often be impossible to achieve with a polymer of low $T_g$. The polymers that most clearly meet these criteria are acrylics or copolymers of vinyl acetate or styrene. These would, without additions of coalescing species, be brittle, forming incoherent films with little adhesion to the substrate at normal application temperatures.

Coalescing solvents allow these polymeric systems to form films at ambient or sub-ambient temperatures. The presence of the coalescing solvent has the following effects.

1 It reduces the total surface energy of the system by reducing polymer surface area.
2 It increases the capillary forces by the controlled evaporation of the water.
3 It reduces the repulsive forces between the polymeric particles.
4 It allows deformation of the particles in contact with each other by effectively lowering the $T_g$ of the polymer.

An emulsion polymer consists of a dispersion of polymeric particles varying in size from 0.05–1 µm, dispersed in an aqueous environment. The coalescing solvent may be found in several different locations depending on their nature. They are classified according to their preferred location in the aqueous system.

Hydrophobic substances such as hydrocarbons will prefer to be within the polymer particle; these are described as A Group coalescents (Figure 5). These tend to be inefficient coalescing agents. Molecules which are more hydrophilic than hydrocarbons tend to be preferentially sited at the particle–aqueous interface, along with the surfactant system. These are described as AB Group coalescents. These exhibit the best efficiency as coalescing solvents.

A : Hydrophobic Environment
B : Boundary Region
C : Hydrophilic Environment

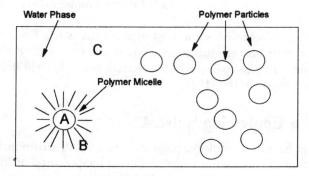

**Figure 5** *Location of species in aqueous dispersions*

The third group is the hydrophobic glycol ethers and similar species. These exhibit good coalescing power, but are partitioned between the polymer particle, the boundary layer and the aqueous phase. More of these are required than AB Group coalescents; they are called ABC Group coalescents. Finally, hydrophilic species such as glycols and the more polar glycol ethers are inefficient coalescing agents, and are more commonly used as freeze–thaw stabilizers. These are described as C Group coalescents.

# 5  Properties of Coalescing Agents

## 5.1  Hydrolytic Stability

A coalescing agent should have good hydrolytic stability (a high degree of resistance to hydrolysis) so that it can be used successfully in both low and high pH latex systems.

## 5.2 Water Solubility

It is desirable for a coalescing aid to have low water miscibility for the following reasons:

a When added to a coating formulation, a coalescing aid with low water miscibility partitions into the polymer phase and softens the polymer; this improves pigment binding and polymer fusion.

b There is less tendency for the evaporation of the coalescing aid to be accelerated by evaporation of the water during the early stages of drying.

c The early water resistance of the coating is not adversely affected.

d When a latex coating is applied to a porous substrate, water immiscible coalescing aids are not lost into the porous substrate along with the water; *i.e.* more coalescing aid will be available to coalesce the coating.

## 5.3 Freezing Point

The freezing point of a coalescing agent should be low (below $-20\,^\circ$C) as materials with a high freezing point may require specialized (therefore more expensive) handling techniques in transport and storage.

## 5.4 Evaporation Rate

The evaporation rate of a coalescing aid should be slow enough to ensure good film formation of the emulsion coating under a wide range of humidity and temperature conditions; however, it should be fast enough to leave the coating film in a reasonable length of time and not cause excessive film softness. The evaporation rate of the coalescing aid should be less than that of water but not so slow that it remains in the film for an extended period of time, thereby causing dirt pickup.

## 5.5 Odour

The odour of a coalescing agent should be minimal. This is especially important for interior coatings applications.

## 5.6 Colour

A coalescing agent should be colourless, to prevent discoloration of the product.

## 5.7 Coalescing Efficiency

The effectiveness of a coalescing agent is based upon the MFFT test.

## 5.8  Incorporation

A good coalescing agent should be capable of addition during any stage of paint manufacture. Any particular coalescing agent may need premixing before addition with varying amounts of the water and surfactant used in the paint stage.

## 5.9  Improvement of Physical Properties

A good coalescing agent can provide a significant improvement in scrub resistance, stain resistance, flexibility and weatherability.

## 5.10  Biodegradability

A good coalescing agent should exhibit acceptably rapid biodegradation.

## 5.11  Safety

The use of the coating system requires that the nature of the coalescing agent should permit use in the domestic environment. The suitability of such a material is therefore obviously dependent on its toxicity, flash point, vapour pressure and VOC classification.

# 6  Comparison of Coalescing Solvents

From the above, it may readily be seen that AB type coalescing solvents give the best performance in use. These are the high $M_r$ esters and ester alcohols. Other species have some of the benefits of these materials but have side effects that render them less suitable. Minimization of the quantity used tends to be a key financial issue, as well as offering the best environmental option. In virtually all cases, the efficiency of the AB group in application is so marked that they have become the industry standards. Several other properties, perhaps less important than efficiency, must then be considered when selecting between these.

These products are usually used in enclosed spaces, where product odour will be vital for acceptance to the consumer. The dibasic esters are virtually odour free. For legislative reasons, a solvent not classified as a VOC will be of definite benefit, given the current attitude to organics in consumer products. Again, the dibasic esters have initial boiling points considerably above the VOC threshold of 250 °C.

The stability of the product towards hydrolysis is also a significant factor in the selection of the optimum coalescing agent. Effectively, this parameter determines the storage stability of the product. Again, dibasic esters, particularly those produced from higher $M_r$ alcohols, appear to have advantage.

The efficiency of the AB group of coalescents towards almost all polymer types has already been mentioned. For styrene–acrylics and vinyl acetate–

veova copolymers, the diesters have a significant edge in performance compared with other members of that group. This advantage is shown in the reduction in amount required to attain a particular MFFT or the actual MFFT for a given addition level.

Chemoxy's Coasol B is a blend of the di-isobutyl esters of three dibasic acids, namely adipic, glutaric and succinic. These acids are produced as a by-product in the production of adipic acid in the manufacture of Nylon 6,6. The ratio of the acids is 15–25% adipic, 50–60% glutaric and 20–30% succinic. Esterification using isobutanol is then performed and the product is isolated following distillation. Coasol B was introduced in the early 1990s as a coalescing agent to the aqueous based coatings industry.

Coasol B has excellent coalescing properties and is compatible with virtually all paint systems (Figure 6). It has very low vapour pressure at ambient temperature, which ensures excellent plasticization throughout the film forming and drying process. Its extremely low odour also significantly improves the general view that aqueous systems have unpleasant odours. This is almost always as a result of the coalescing solvent. Its resistance to hydrolysis allows its use in a wide variety of formulations, including those that require adjustment of pH to basic conditions.

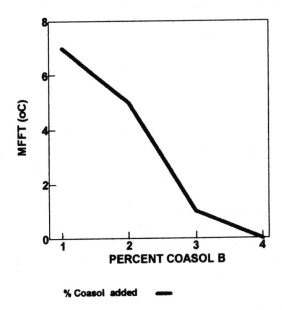

% Coasol added ➡

**Figure 6** *Evaluation with EMULTX VV 536 acetate/veova copolymer dispersion*

In addition, it rapidly biodegrades, both aerobically and anaerobically, since, as an ester, it is readily attacked by ubiquitous bio-species. Finally, its toxicological profile is virtually benign, so in use it can be handled with confidence in most working environments. Its properties are summarized in Table 1.

**Table 1** *Properties of Coasol B*

| Physical characteristic | Coasol B | Comments |
|---|---|---|
| Boiling point | > 275 °C | Not a VOC |
| Evaporation rate | < 0.01 (Butyl acetate = 1.0) | Slow, allowing excellent coalescence |
| Freezing point | < − 55 °C | No freeze–thaw issue |
| Water solubility | 600 ppm | |
| Hydrolytic stability | Very good | No hydrolysis in normal use. |
| Colour | 5 Hazen units | Imparts no colour to coatings |
| Coalescing efficiency | Excellent for most polymer systems | |
| Biodegradability | 80% in 28 days | |
| | | Biodegradable |
| Odour | None discernible | |
| Toxicology | Oral $LD_{50}$ (Rat) >16000 mg $kg^{-1}$ | Essentially non-toxic |

# 7  Recent Advances in Diester Coalescing Solvents

More recently, Chemoxy has been evaluating a range of esters as Coalescing agents in paint formulations. Firstly, we have prepared a formulation similar to Coasol B using a higher proportion of adipic acid to see how this affects coalescing performance. This new product is called Coasol A (on account of the higher adipate content). The ratios of acids in this product are 35–40% adipic, 45–55% glutaric, and 10–15% succinic. The vapour pressure at room temperature of this material is lower than that of Coasol B, and its boiling range is also higher.

We have also prepared the di-isopropyl esters of the higher adipic content stream. This has a vapour pressure similar to that of Coasol B, but is slightly more water-soluble. Finally, we have manufactured di-isopropyl adipate, which has the highest boiling point, the lowest vapour pressure and the lowest water solubility of all of this range of products. These preparations were undertaken to add to our repertoire of products to suit the diverse require-ments of the formulators of aqueous based systems.

Comparative formulations using Coasol B, di-isopropyl AGS and di-isopropyl adipate in comparison with a monoester of pentane diol were carried out to determine the performance of each coalescing agent. In virtually all cases, the dibasic esters gave a significant improvement in efficiency in reducing the MFFT for a given quantity of additive. Figures 7–12 below show the performance of three diesters and pentane diol ester in standard Harco Polymer dispersions. We at Chemoxy are attempting to offer a tailor made solution to each individual polymer system employed in the development of aqueous based systems. We believe that the dibasic esters of the AGS acids group offer the opportunity for fine tuning, with the added advantage of low odour, low toxicity and 'excellent' VOC status.

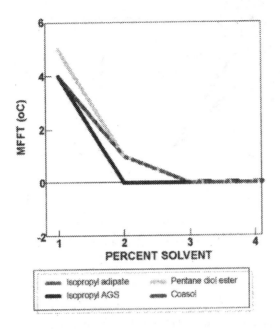

**Figure 7** *Evaluation with MOWILITH DM21, a vinyl acetate/veova copolymer dispersion*

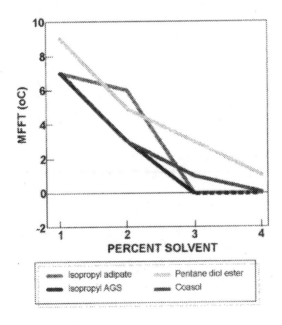

**Figure 8** *Evaluation with VIKING 5488, a hard styrene/acrylic ester copolymer dispersion*

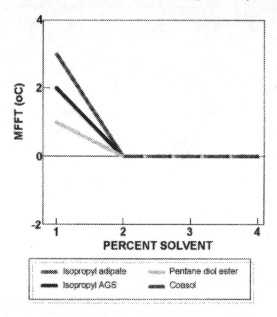

**Figure 9** *Evaluation with MOWILITH DM123, a cellulose ether stabilized vinyl acetate/ vinyl chloride terpolymer dispersion*

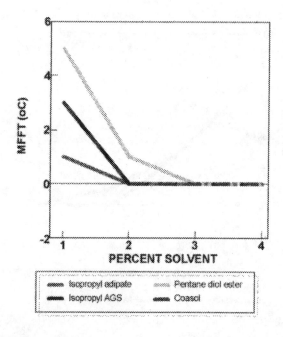

**Figure 10** *Evaluation with REVACRYL 239, a styrene/acrylic ester copolymer dispersion*

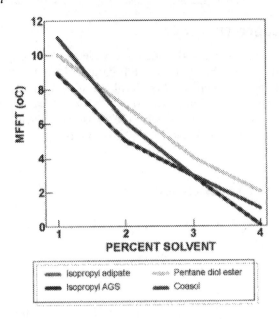

**Figure 11** *Evaluation with EMULTEX VV573, a cellulose ether stabilized vinyl acetate/ veova 10 copolymer dispersion*

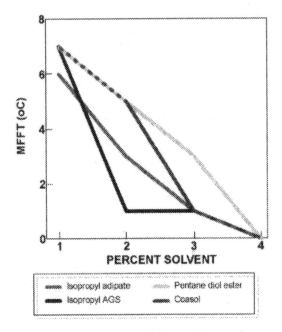

**Figure 12** *Evaluation with EMULTEX VV536, a vinyl acetate/veova copolymer dispersion*

## Acknowledgements

I would like to acknowledge with much gratitude the help given by Mr R J Foster of Harco in assembling the MFFT data for this paper. I must also thank my colleagues at Chemoxy, Ms Carol White, who assembled much of the data used, and Miss Tracy McGough, who helped me produce the diagrams. Finally, I must acknowledge the assistance given by Mr T J P Thomas, who has acted as a consultant to Chemoxy International in this whole area.

## Appendix – Classification of Coalescing Solvents

| Coalescent type | Type of species | Examples | Comments |
|---|---|---|---|
| Type A | Hydrocarbons | White spirit | |
| Type AB | Diesters | 1. DBE Dimethyl esters<br>2. DBE Di-isobutyl esters<br>3. Di-isobutyl adipate<br>4. Di-isopropyl adipate<br>5. Dibutyl phthalate | Estasol, Du Pont DBEs<br>Coasol, Lusolvan<br>Chemoxy new products |
| Type AB | Ester alcohols | Diol monoesters | Texanol |
| Type ABC | Glycol esters & Glycol ester ethers | 1. PGDA<br>2. Butyldiglycol acetate | |
| Type ABC | Ether alcohols & diethers | 1. PnBs<br>2. 2-Butoxyethanol<br>3. MPG diethers | Dow products<br>BASF and others<br>Proglides and glymes |
| Type C | Glycols | 1. DEG<br>2. DPG<br>3. TEG | |

# Application of Biocides in Waterborne Coatings

Klaus Weber

SCHÜLKE & MAYR GMBH, D-22840 NORDERSTEDT, GERMANY

## 1 Introduction

### 1.1 Microorganisms

Microorganisms can be found in nearly every environment. There are even species living in hot springs or under anaerobic conditions. Few requirements are necessary for microbial growth. Certainly there must be a source of carbon or nitrogen, mineral salts, sometimes the energy of light and, of course, water. The absence of water limits every form of life. Many species develop a sporeform to get through times of dryness.

Water based coatings fulfil all these requirements in a perfect way. Nearly every raw material in coating systems can be a nutrient for microorganisms. Water is available in the containers and also on the dry film by means of condensation or rain. Many different microorganisms such as bacteria, yeasts, mould fungi and algae can be found in contaminated coating systems.

### 1.2 Signs of Contamination

A microbial contamination results in deterioration and changes in the macroscopic properties of the product. Typical signs of microbial growth are

- Changes in viscosity
- Changes in pH value
- Discolorations on the surface
- Coagulation
- Gas formation
- Foul odour
- Visible growth
- Cracks in the surface of the dry film

## 1.3 Sources of Contamination

To bring microorganisms into the products is quite easy. Many raw materials such as fillers are often highly contaminated. In practice unpreserved stock solutions and, at least, the water used in the process play an important role. Deficient industrial hygiene and even the air can add to the problem.

A strict monitoring system for all these details should be implemented. Rapid microbial colonization of every surface takes place whenever the level of moisture is sufficient for the growth of fungi, algae and mosses. Spores of fungi are brought in by air, and other organisms follow. This applies to both interiors and exteriors. Under favourable conditions of moisture, temperature and supply of nutrients spores enter the vegetative phase. In this primary invasion, fungi feed on the coating ingredients, which are thus either damaged or in some cases destroyed. A secondary invasion, in which the fungi grow on a layer of dust and dirt, does not necessarily damage the coatings.

This kind of microbial contamination cannot be avoided. Spores are present at different levels everywhere in the environment. Their increase can only be inhibited to protect surfaces from growth that becomes visible by discolorations.

## 1.4 Preservation

There is no doubt about the necessity of preservation in water-based coating systems. During the production process, storage and transport in the closed containers the in-can preservation protects the product from microbial attack. It must be pointed out that biocides for in-can preservation serve only to inhibit microbial growth in the finished product and not to decontaminate highly polluted products throughout the production process. Because the action of many preservatives causes a consumption of the active ingredients, commonly used concentrations would be too low to fulfil this purpose. Higher concentrations would be necessary to guarantee safe products during storage and transport. Since preservatives are cost-intensive, overdosing of ingredients should be avoided by careful control of raw materials and production hygiene.

Underdosing is also critical as a small loss of concentration caused by sorption, chemical reactions, or whatever reason, will limit the safety. For the use of biocides in preservation there is a simple rule to follow

- Use as little as possible
- Use as much as technically necessary

However, the dry film also has to be protected by efficient biocides to avoid growth of fungi and algae on the surface. These film preservatives do not have to show their effect during the relatively short period of production, storage and transport, but rather for a very long time on different surfaces. For this purpose a different variety of active ingredients is used.

# 2 In-can Preservation

## 2.1 Requirements for Active Agents

Many active agents of different chemical structures are used as preservatives. For in-can preservation there are a number of requirements:

- Broad spectrum of effect against bacteria, yeasts and mould fungi
- Rapid onset of effect (biocidal effect on microorganisms from raw materials)
- Long lasting effect (preservation)
- Water-solubility in the concentrations used
- Head-space effectiveness
- Stability over a wide pH range
- Good compatibility with all ingredients and materials
- No discolorations in container and film
- Easy incorporation, dosing and handling
- No effect on other than biological properties
- Acceptable risk for man, animals and the environment during use and on disposal
- High degree of cost-effectiveness

Since today there is no particular substance combining all these properties in a practicable way, combinations and synergistic mixtures make up acceptable compromise solutions.

## 2.2 Chemistry of the Active Agents

Organic compounds of a wide variety of different structures are used as active agents in biocidal preparations. The following list shows the principal compounds used for in-can preservation.

In the past, heavy metal compounds and halogenated phenol derivatives of excellent efficacy were used in preservation. However, they no longer meet today's requirements for safety and environmental acceptability. For this reason they will not be discussed here.

Despite all public debate, formaldehyde-releasing compounds are widely used. The most common examples are given below:

O-formals: reaction products from formaldehyde and alcohols or glycols such as isopropanol and benzyl alcohol or ethylene glycol and its ethers (Figure 1).

N-formals: reaction products from formaldehyde and amides such as chloroacetamide or urea (Figure 2).

In addition, reaction products from amines or aminoalcohols that form mixed N,O-formals from ethanolamine or isopropanolamine (Figure 3).

Other formaldehyde-releasing compounds include bronopol.

Formaldehyde-free agents are available as active oxygen compounds, such as *tert*-butyl hydroperoxide.

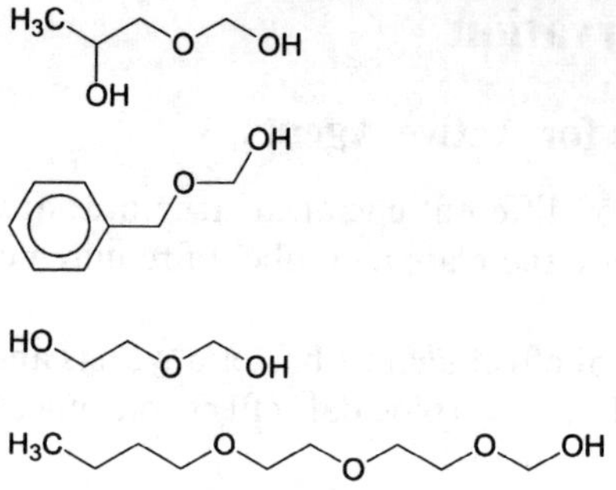

**Figure 1** *O-Formals*

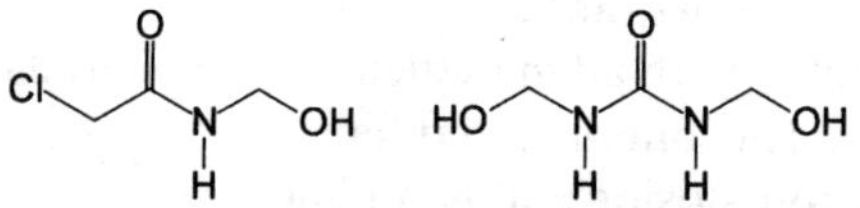

**Figure 2** *N-Formals*

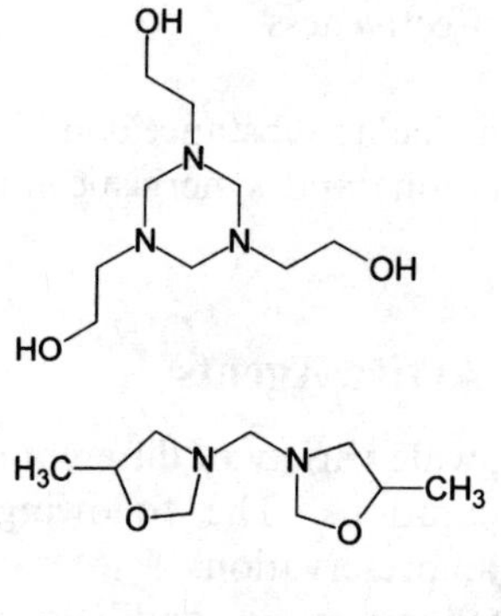

**Figure 3** *N,O-Formals*

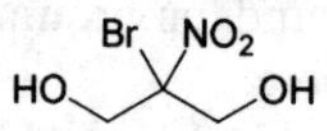

Bronopol

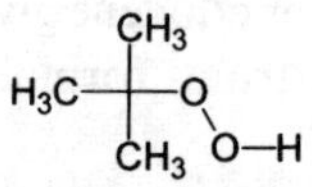

tert-Butyl hydroperoxide

Another class of compounds used as in-can preservatives are isothiazolinone derivatives (Figure 4).

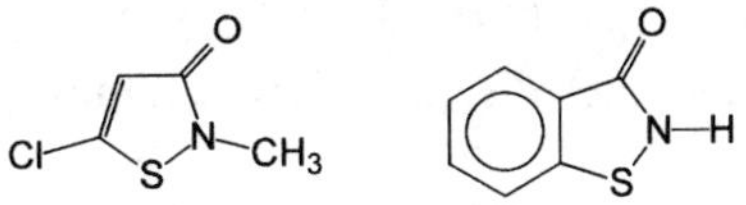

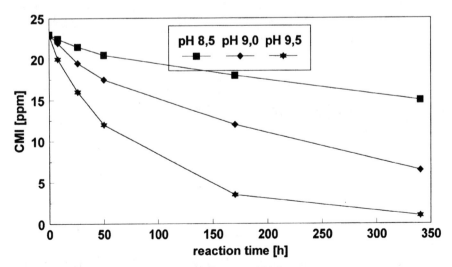

**Figure 4** *Isothiazolinones*

## 2.3 Stability of Assorted Active Agents

The effect of many active agents is based on a chemical reaction. Because of this reaction actives are consumed during their use. Not just this process but also other influences (*e.g.* extreme temperatures and/or pH values, reducing or oxidizing conditions and strong nucleophilic/electrophilic ingredients) lead to losses in concentration. These special instabilities are well known from many active agents. Since many end users look for alternative actives to formaldehyde and formaldehyde-releasing substances, isothiazolinones have become more and more widespread in preservation. But these particular compounds, which are normally used in mixtures of chloromethylisothiazo-linone (**A**) and methylisothiazolinone (**B**) are not very stable at high pH values.

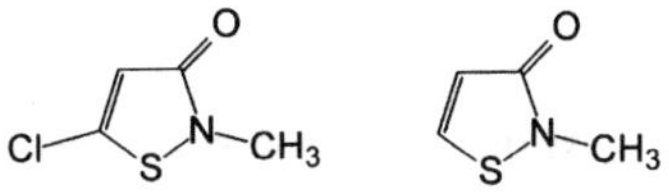

Under buffered alkaline aqueous conditions the more active chloromethyl-isothiazolinone (CMI) is destroyed much faster than the methylisothiazoli-none, which remains unchanged during the experiment (Figure 5).

**Figure 5** *Stability of chlorometylisothiazolinone at 30 °C*

Alkaline systems containing CMI are also very temperature sensitive. An increase of just 10 °C accelerates the degradation significantly (Figure 6).

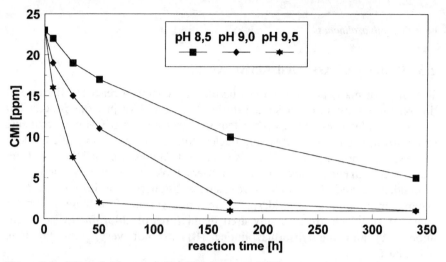

**Figure 6** *Stability of chloromethylisothiazolinone at 40 °C*

*2.3.1 Mechanism.* The degradation of chloromethylisothiazolinone (**1**) starts with the hydrolysis of the vinylic chlorine to the enolate **3**. Afterwards the active centre of the resulting cyclic thioester derivative **4** is attacked by hydroxide and ring opening occurs. Finally the open-chained **5** loses sulfur (which is visible as a turbidity in aqueous solutions) to build the ineffective malonic amide **6** (Scheme 1).

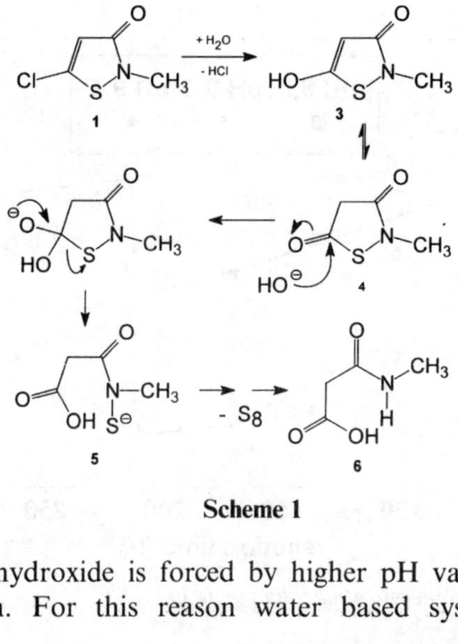

**Scheme 1**

The attack of hydroxide is forced by higher pH values to accelerate the whole mechanism. For this reason water based systems preserved with isothiazolinones must be protected from heat and high pH values. Benzisothiazolinone is more stable because of the stabilizing effect of the aromatic ring system.

# 3 Film Preservation

## 3.1 Requirements for Active Agents

As mentioned above there are big differences between film preservatives and in-can preservatives. Active agents for film preservation are subject to a number of further requirements:

- Effect against fungi and algae
- Long lasting effect
- Extremely low water solubility
- Low vapour pressure
- Stability in the presence of UV light
- Stability over a wide pH range
- Good compatibility with all ingredients and materials
- No discolorations in container and film
- Easy incorporation, dosing and handling
- No effect on other than biological properties
- Acceptable risk for man, animal and the environment during use and on disposal
- High degree of cost-effectiveness

All these requirements have not yet been fulfilled by a single active agent. As with in-can preservatives, combinations of substances have to be used.

## 3.2 Chemistry of Active Agents

Many substances have been used as film preservatives in the past. For example, organic heavy metal compounds are still very successful in anti-fouling paints. Apart from this application preparations of this kind should no longer be used for environmental reasons. Because of their stability and their long lasting effectiveness as biocides, active agents for film preservation are always a subject of discussion. The sometimes-underestimated value of these compounds lies in the protection of huge amounts of material from destruction by the attack of microorganisms.

The following is a summary of the most common substances for film preservation.

*3.2.1 Urea Derivatives.* $N'$-(3,4-dichlorophenyl)-$N,N$-dimethylurea (Diuron) is an example of a very selective biocide. It is nearly ineffective against fungi or bacteria and shows very low mammalian toxicity.

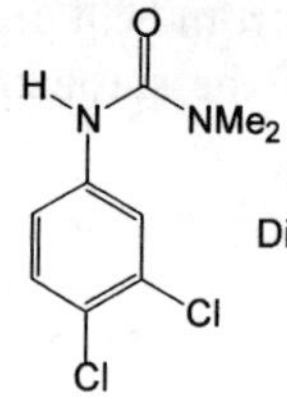

Diuron

Diuron is poorly soluble in water. However, it inhibits photosynthesis and is therefore suitable for algicidal finishing.

*3.2.2 Isothiazolinones.* The activated S–N bond of isothiazolinones is able to react with nucleophilic compounds. On the one hand it makes some isothiazolinones very unstable as shown above. But it also enables these compounds to react with nucleophilic amino- or thiol-groups located on the cell wall of microorganisms. This reaction leads to damage of the organism.

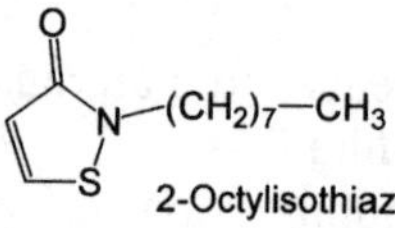

2-Octylisothiazolinone

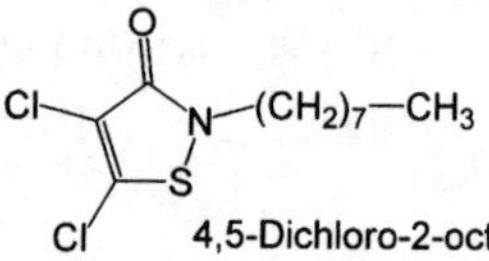

4,5-Dichloro-2-octylisothiazolinone

The reaction is also irreversible and therefore the biocide is slowly consumed. For a suitable long-term protection there must be a sufficient amount of active ingredients incorporated in the surface. The well-known instability at high pH values also limits the application of isothiazolinones.

*3.2.3 Dithiocarbamate Derivatives.* This class of compounds is suitable for a broad range of applications. Due to their properties they can be used as fungicides in almost every kind of coating system. Bis(dimethylthiocarbamoyl)-disulfide (Thiram) and zinc-*N*-dimethyldithiocarbamate (Ziram) are stable to light, nearly insoluble in water and not volatile.

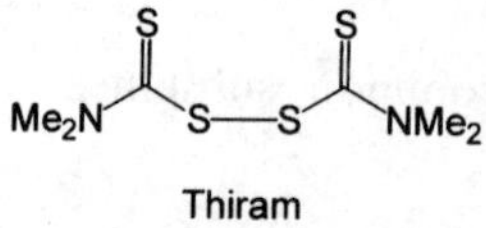

Thiram

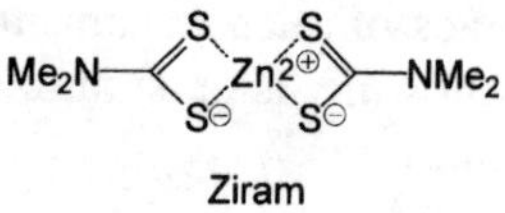

Ziram

*3.2.4 Benzimidazole Derivatives.* Benzimidazole derivatives are well known as fungicides. 1*H*-Benzimidazol-2-ylcarbamic acid methyl ester (Carbendazim) inhibits the DNA-synthesis of cells at concentrations of less than 1 ppm. Despite its low toxicity in mammals it is suspected to have mutagenic properties. Carbendazim is stable and effective at a pH range from 2 to 12. Because of certain gaps in the effectiveness it must be combined with other fungicides.

2-(4-Thiazolyl)-benzimidazole (Thiabendazole) is almost as effective as

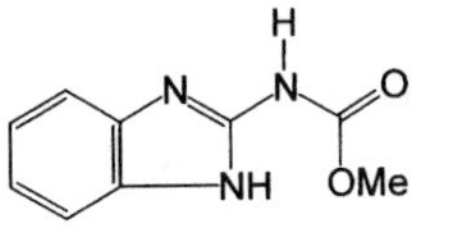

Carbendazim

carbendazim. Due to its comparable higher water solubility, particularly at low pH values (3.84% at pH 2.2!) it is suitable only for special applications.

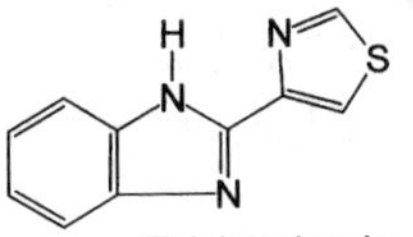

Thiabendazole

*3.2.5 Triazine Derivatives.* Some triazines show excellent algicidal properties. *N'*-*tert*-butyl-*N*-cyclopropyl-6-(methylthio)-1,3,5-triazine-2,4-diamine (Irgarol 1051) additionally inhibits the growth of mussels which makes it suitable as an anti-fouling component for ship-bottom paints.

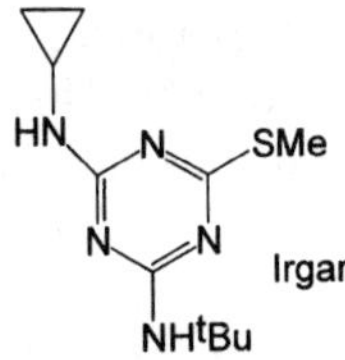

Irgarol 1051

*3.2.6 Benzothiazole Derivatives.* 2-(Thiocyanomethylthio)-benzothiazole (TCMTB) is extensively used in the leather industry. Although TCMTB is relatively sensitive to hydrolysis its broad fungicidal activity is ideal for closing the gaps in the effectiveness of other active agents.

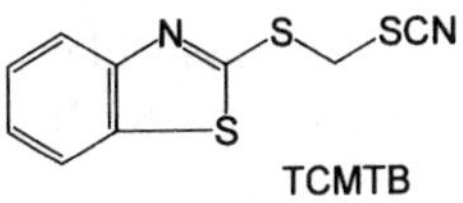

TCMTB

*3.2.7 Carbamic Acid Derivatives.* Known from wood protection, the fungicide 3-iodo-2-propynyl butylcarbamate (IPBC) is also registered for cosmetic preservation. Supplementary to its fungicidal activity it shows a weak effect on algae. IPBC tends to be leached out due to its relatively good water solubility (0.015%). Like all iodine compounds it may cause discoloration.

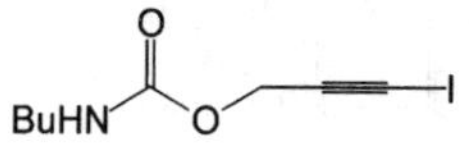

IPBC

*3.2.8 Thiophthalimide Derivatives.* Suitable for special applications, this is another class of substances with an activated S–N bond as reactive centre. The fungicides *N*-[(trichloromethyl)thio]phthalimide (Folpet) and *N*-[(dichlorofluoromethyl)thio]-phthalimide (Fluorofolpet) are interesting components for fungistatic paints because of their low water solubility. Unfortunately both of them are easily degraded by hydrolysis. The more stable, Folpet, can be used in special dispersion paints; activated by an additional fluorine atom, Fluorofolpet is stable only in alkyd paints.

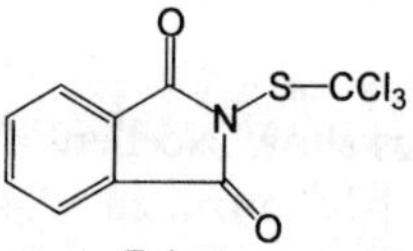

Folpet 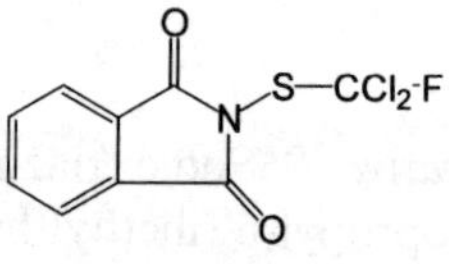                    Fluorofolpet

*3.2.9 Sulfenic Acid Derivatives.* Comparable structures can be found in the sulfenic acid amides 1,1-dichloro-*N*-[(dimethylamino)sulfonyl]-1-fluoro-*N*-phenylmethansulfenamide (Dichlofluanid/Preventol A4S) and 1,1-dichloro-*N*-[(dimethylamino)sulfonyl]-1-fluoro-*N*-(*p*-tolyl)methansulfenamide (Tolylfluanid/Preventol A5).

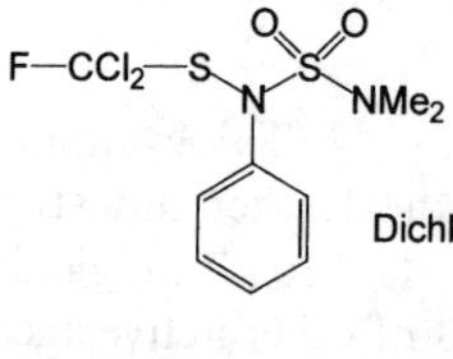

Dichlofluanid

Both compounds are excellent fungicides but undergo easy hydrolysis. Dichlofluanid is one of the leading fungicides used in wood paints and wood stains. With the introduction of the BPD it will be replaced by Tolylfluanid.

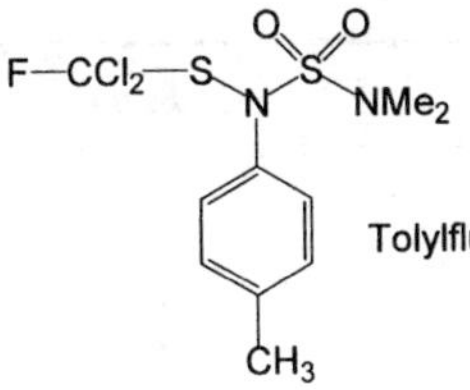

Tolylfluanid

### 3.2.10 Sulfone Derivatives.

*3.2.10 Sulfone Derivatives.* 1-[(Diiodomethyl)sulfonyl]toluene (Amical) is a fungicidal compound with low water solubility and a broad-spectrum activity. The mechanism of activity is based on a transfer of iodine to nucleophilic components of the microbial cells. This activated iodine compound tends to cause discoloration.

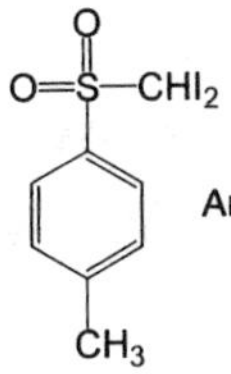

Amical

*3.2.11 Pyridine-N-oxide Derivatives.* Because of its low solubility in water (15 ppm) and broad spectrum of activity zinc-bis(2-pyridinethiol-1-oxide) (Zinc Pyrithione/Zinc Omadine) is a very interesting fungicidal and respectively algicidal compound.

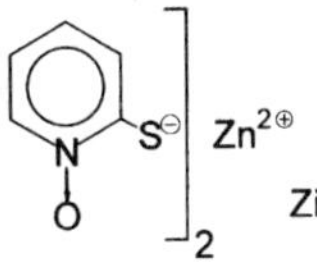

Zinc Pyrithione

The molecule is stable to moderately high pH values and not even higher temperatures lead to significant decomposition. Attention must be paid to oxidizing components such as peroxides or active chlorine compounds that transform the compound into the inactive sulfonate. High levels of iron ions from the water used in production may lead to intensely coloured iron complexes of pyridinethione.

## 3.3 Spectrum of Efficacy

Table 1 summarizes the efficacy spectrum of all mentioned active agents in film conservation.

**Table 1** *Spectrum of efficacy*

| Class of actives | Fungi | Algai |
|---|---|---|
| Organic metal compounds | + | + |
| Urea derivatives | − | + |
| Isothiazolinones | + | + |
| Dithiocarbamate derivatives | + | − |
| Benzimidazole derivatives | + | − |
| Triazine derivatives | − | + |
| Benzothiazole derivatives | + | (+) |
| Carbamic acid derivatives | + | (+) |
| Thiophthalimide derivatives | + | (+) |
| Sulfenic acid derivatives | + | (+) |
| Sulfone derivatives | + | − |
| Pyridine-*N*-oxide derivatives | + | (+) |

( ): with limitations.

## 3.4 Stability and Solubility of Actives in Film Preservation

Table 2 summarizes the stability and solubility of the active agents.

**Table 2** *Stability and solubility*

| Class of actives | Stability | | Solubility | | |
|---|---|---|---|---|---|
| | alkaline aqueous systems | light | H₂O | Organic solvent | |
| | | | | Polar | Non-polar |
| Organic metal compounds | + | (+) | − | + | + |
| Urea derivatives | + | + | − | + | + |
| Isothiazolinones | (+) | + | − | + | (+) |
| Dithiocarbamate derivatives | (+) | | − | (−) | (−) |
| Benzimidazole derivatives | + | − | − | + | |
| Triazine derivatives | + | + | − | + | + |
| Benzothiazole derivatives | (+) | (+) | − | (+) | (+) |
| Carbamic acid derivatives | (+) | (+) | − | + | (+) |
| Thiophthalimide derivatives | − | (+) | − | (+) | (+) |
| Sulfenic acid derivatives | − | (+) | − | + | (+) |
| Sulfone derivatives | + | − | − | + | (+) |
| Pyridine-*N*-oxide derivatives | + | (+) | − | − | − |

( ): with limitations.

## 4 Conclusions

Contamination of products and surfaces is inevitable. A certain amount of contamination can be avoided during the production process by following the rules of good production hygiene and an efficient microbiological control system for raw materials. Subsequent microbial contamination and growth during storage, transport and use must be kept under control by using the right preservative in the right concentration.

There are many different active agents used for in-can and film preservation. Preservatives for particular applications have to be evaluated according to different criteria:

- Spectrum of efficiency
- Dosage
- Safety for man, animals and environment

In practice these requirements can only be fulfilled by combinations of different active agents. Cost-effectiveness must be kept in mind during purchase, storage and the production process. From this point of view it must be concluded that it is more advantageous to work with manufactured commercial products instead of pure chemicals.

# A Novel Dispersant for Waterborne Resin-Free Pigment Concentrates

Janos Hajas and Andreas Ahrens

BYK-CHEMIE GMBH, ABELSTRASSE 14, D-46483 WESELI, GERMANY

## 1 Introduction

## 2 Characteristics of the Novel Wetting and Dispersing Additive

The new polymeric wetting and dispersing additive is produced in commercial quantities, and is marketed as the product identified in the footnote.[*]

Technical data is given in Table 1.

**Table 1** *Technical characteristics of the novel wetting and dispersing additive*

| | |
|---|---|
| Chemical composition | Solution of a high $M_r$ block copolymer with pigment affinic groups |
| Functionality | Carboxylic |
| Acid value (mg KOH Ig) | 10 |
| Refractive index | 1.40 |
| Delivery form | 40% in water (amine-free, co-solvent-free) |
| Pigment stabilization | 1. By anionic charge |
| | 2. By steric hindrance |

## 3 Comparison with Conventional Wetting and Dispersing Additives

Comparative studies utilizing white and black pigmentation (titanium dioxide and carbon black) were performed in various resin systems such as acrylic and

[*] Disperbyk-190.

styrene–acrylic emulsion polymers, water-soluble air drying resin systems and hybrid baking resin combinations. The stability of each formulation was tested (phase separation, pigment settling, visible flocculation), along with gloss and water resistance of the resultant coating films.

## 3.1 Results in Emulsion Polymers

A white-pigmented, glossy topcoat was formulated with an alkylphenol ethoxylate wetting agent, an ammonium polyacrylate wetting and dispersing additive, and the novel wetting and dispersing additive (at 2.5% active substance on the pigment).

Utilizing resin-free pigment concentrate technology (very high pigmentation level – 70% titanium dioxide in the millbase), the pigment was dispersed in water with each wetting and dispersing additive. Letdown was performed by utilizing the emulsion polymer. Results are shown in Table 2.

**Table 2** *Properties of white glossy emulsion paints containing various wetting agents and dispersants*

|  | Stability | Gloss | Water resistance |
|---|---|---|---|
| Wetting agent | Moderate | Good | Poor |
| Polyacrylate dispersant | Excellent | Good | Good |
| Novel wetting and dispersing additive | Excellent | Good | Good |

As a second pigmentation, a black coating was formulated with carbon black (P.B.7). Since carbon black has a very high surface area (approximately 40–50 times higher than the specific surface area of common titanium dioxide), the additive dosage was increased to 70% active substance on the pigment. The pigmentation level in the concentrate was 20%, and in the final coating 2.5%. Results are shown in Table 3.

**Table 3** *Properties of black glossy emulsion paints containing various wetting agents and dispersants*

|  | Stability | Gloss | Water resistance |
|---|---|---|---|
| Wetting agent | Poor | Moderate | Poor |
| Polyacrylate dispersant | Poor | Poor | Good |
| Novel wetting and dispersing additive | Good | Good | Good |

Similar results can be obtained with most organic pigments; evaluations have included Pigment Yellow 74, Pigment Red 122, Pigment Violet 23, *etc*. This suggests that conventional wetting agents and dispersants can be used in emulsion paints only for dispersion of inorganic pigments and extenders.

## 3.2 Results in Water-soluble Systems

A water soluble alkyd air drying alkyd resin was employed. Pigmentation and preparation of batches was the same as in Section 3.1. Some results are displayed in Tables 4 and 5.

**Table 4** *Properties of white glossy water-soluble paints containing various wetting agents and dispersants*

|  | Stability | Gloss | Water resistance |
|---|---|---|---|
| Wetting agent | Good | Good | Poor |
| Polyacrylate dispersant | Poor | Poor | Good |
| Novel wetting and dispersion additive | Excellent | Excellent | Good |

**Table 5** *Properties of black glossy water-soluble paints containing various wetting agents and dispersants*

|  | Stability | Gloss | Water resistance |
|---|---|---|---|
| Wetting agent | Good | Poor | Poor |
| Polyacrylate dispersant | Poor | Poor | Poor |
| Novel wetting and dispersing additive | Excellent | Excellent | Good |

Very similar results have been obtained with the pigments listed in Section 3.1. These facts suggest that conventional dispersants (products that provide only charge stabilization) cannot be used in water-soluble formulations for either inorganic pigments or for carbon black.

## 3.3 Results in Hybrid Systems

A waterborne colloidal acrylic emulsion polymer was used as a water-soluble melamine resin, in combination with a resin-free mill-base (the same as in Section 3.1), to formulate baking topcoats. Formulations with titanium dioxide and carbon black were produced accordingly; the results can be seen in Tables 6 and 7.

**Table 6** *Properties of white glossy hybrid baking coatings containing various wetting agents and dispersants*

|  | Stability | Gloss | Water resistance |
|---|---|---|---|
| Wetting agent | Moderate | Good | Poor |
| Polyacrylate dispersant | Good | Moderate | Poor |
| Novel wetting and dispersing additive | Excellent | Excellent | Good |

**Table 7** *Properties of black glossy hybrid baking coatings containing various wetting agents and dispersants*

|  | Stability | Gloss | Water resistance |
| --- | --- | --- | --- |
| Wetting agent | Poor | Poor | Poor |
| Polyacrylate dispersant | Poor | Poor | Poor |
| Novel wetting and dispersing additive | Excellent | Excellent | Good |

# 4 Advantages of the New Wetting and Dispersing Additive Technology

Based on the results in emulsion polymers, water-soluble systems and hybrid systems, only the novel wetting and dispersing additive was able to provide superior dispersion stability, high gloss and good water resistance in *all* resin systems with various pigments.

An additional benefit was the very low mill-base and pigment concentrate viscosity. Conventional wetting agents only function well with water-soluble resins and inorganic pigmentation. Conventional dispersants are suitable for inorganic pigments, emulsion polymers and hybrid systems, but they completely fail in water-soluble resins.

The excellent results obtained with this novel wetting and dispersing additive can be explained by its special chemical structure – a structure that provides both *charge stabilization* and *steric stabilization* of the pigment particles.

# 5 Performance Characteristics: Grinding in Resin or Resin-free?

## 5.1 Full Shade Pigmentation

Emulsion paints are usually dispersed without a grinding resin but with wetting and dispersing additives and, if necessary, additionally with selected wetting agents in water, followed by letdown with the emulsion polymer. This technique is widely used with inorganic pigments and extenders, but it has limitations with carbon blacks and with organic pigments.

The traditional method of grinding in water-soluble and hybrid systems is to use a low viscosity grinding resin. The novel wetting and dispersing additive can also be used with 'grind-in-resin' technologies, but dispersion results are, in most cases, worse than by those obtained by producing resin-free grindings.

We explain this by the fact that optimal additive adsorption can only be achieved in resin-free grindings because of the absence of competitive adsorption from the resin molecules. Of course, resin interaction cannot usually provide the same degree of stabilization effects as the highly polymeric molecules of the novel wetting and dispersing additive.

A comparison of resin-free and grinding-resin-containing formulations is shown in Table 8. The characteristics of the finished paints (base: hybrid resin combination) are compared in Figure 1.

**Table 8** *Pigment concentrate formulations with the novel wetting and dispersing additive (resin-free and resin-containing white and black systems)*

|  | Resin-free | | Resin-containing | |
|  | White | Black | White | Black |
|---|---|---|---|---|
| Water | 20.2 | 44.0 | 11.7 | 34.0 |
| Novel dispersant (Disperbyk~ – 190) | 8.8 | 35.0 | 5.3 | 25.0 |
| Defoamer | 1.0 | 1.0 | 1.0 | 1.0 |
| Pigment | 70.0 | 20.0 | 70.0 | 20.0 |
| Acrylic grinding resin (pH 8.5) |  | – | 12.0 | 20.0 |
|  | 100.0 | 100.0 | 100.0 | 100.0 |

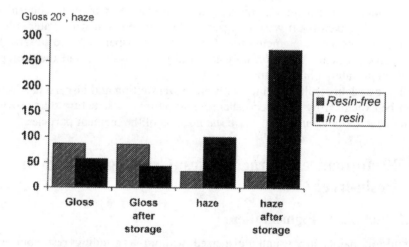

**Figure 1**  *Performance characteristics of hybrid baking coatings made from resin-free and resin-containing pigment concentrates using the novel wetting and dispersing additive*

## 5.2 Mixed Colours

By mixing resin-free and resin-containing pigment concentrates to a grey colour, resin-free grindings can also provide a significant advantage over the pigment concentrates formulated only with grinding resin (Table 9). In this comparative study the final grey colour was achieved by using 90% white plus 10% black pigment concentrate. Letdown was performed in a hybrid-type baking coating.

**Table 9** *Performance characteristics of resin-free and resin-containing pigment concentrates in a grey shade (cross-check)*

| | Resin-free white | | Resin-containing white | |
| | + Resin-free black | + Resin -containing black | + Resin-free black | + Resin -containing black |
|---|---|---|---|---|
| Gloss (200) | 87 | 80 | 67 | 60 |
| DeltaEofrub-up | 0.2 | 0.9 | 1.1 | 2.2 |

Sub-optimal results in resin-containing grinds can be also caused by the alkaline pH of the grinding resin solution, which is not optimal for adsorption of the novel polymeric wetting and dispersing additive. Best results, especially with inorganic pigments, are generally obtained in 'acidic grindings' at slightly acidic pHs (approximately 5.5).

## 5.3 Durability *vs.* Grinding Technique

Surprisingly, resin-free grinding results in better UV resistance and water resistance of the coating than grinding in a resin. Table 10 overviews the durability characteristics of a hybrid-type baking coating (resin system different from those used in 3.3).

**Table 10** *Durability properties of white coatings vs. grinding technique*

| | With the novel wetting and dispersing additive | | No dispersant (control) |
| | Resin-free grinding | Grinding in resin | |
|---|---|---|---|
| Initial gloss (20 °C) | 86 | 80 | 75 |
| Gloss after 500 h QUV (20 °C) | 75 | 67 | 66 |
| Water resistance after 5 days (0 = no blistering, 5 = full of blisters) | 1 | 3 | 1 |

## 6 Applications

The novel wetting and dispersing additive has been designed to produce low viscosity, non-VOC, resin-free pigment concentrates in water, free of any organic co-solvents and amines. Excellent results have been obtained with inorganic pigments (titanium dioxide, red and yellow iron oxide, *etc.*) and organic pigments and carbon blacks, including very high quality pigments such as DPP-BO (Pigment Red 254).

The pigment concentrate formulations display excellent compatibility and letdown characteristics with a wide range of resins, including acrylic emulsion polymers, alkyd emulsions, water soluble alkyds, epoxy esters, polyesters, polyurethanes, acrylic and melamine resins. Furthermore, excellent results with polyester and polyurethane colloidal emulsions and in hybrids are obtained in acrylic systems.

For a better storage stability, it is sometimes necessary to modify the pigment concentrate formulation as follows. For inorganic pigmentation: +0.2–0.4% [anti-settling additive (fumed silica)]. For all colours, if microbial contamination of the pigment concentrate cannot be excluded: +0.05–0.2% (biocide).

As already demonstrated, by utilizing the novel polymeric wetting and dispersing additive in coating formulations, the best results can be obtained with resin-free grindings. In printing ink formulations, however, grindings in suitable grinding resins can also be recommended for several applications. A brief list of various existing applications is given in Table 11.

**Table 11** *Application areas of the novel polymeric pigment wetting and dispersing additive*

| Coating applications (Pigment concentrates for) | Printing ink applications (Pigment concentrates for) |
|---|---|
| General industrial baking systems (Acrylic/melamine, Polyester/melamine) | Flexographic inks for packaging (paper, plastic coated paper, plastic films) |
| General industrial air drying systems (Alkyd emulsions, polyurethane dispersions, water-soluble alkyds) | Gravure inks for packaging (paper, plastic coated paper, plastic films) |
| Automotive coatings (Polyester/melamine baking topcoats, two-pack waterborne polyurethane refinish systems) | Silk screen inks for general and specific applications (paper, plastic, textile) |
| Premium universal waterborne factory colorant lines (from emulsion paints to hybrid and water-soluble systems) | |

# Recovery of Pigment and Additives from Waste Paper Coating Formulations

D.R. Skuse,[1] J.C. Husband,[1] J.S. Phipps,[1] D.C. Payton,[1] J.A. Purdey,[1] M.T. Rundle,[1] and O. Toivonen[2]

[1]CENTRAL RESEARCH LABORATORY, ECC INTERNATIONAL, JOHN KEAY HOUSE, ST. AUSTELL, CORNWALL PL25 4DJ, UK
[2]AMERICAN PAPER GROUP, ECC INTERNATIONAL, 300 MANSELL RD., ROSWELL, ATLANTA, GA, USA

## 1 Introduction

Paper coating formulations, or colours, contain a number of raw materials, including pigment, latex, dispersants, optical brightening agents, thickeners, lubricants, and cross-linkers. Some of these materials are expensive. Inevitably, some colour is lost during web breaks, grade changes and wash-ups. Paper mills producing light weight coated mechanical (LWC) paper, characterized by long runs and few grade changes, are said to lose 2–6 wt.% of their colour. On the other hand, mills producing speciality grades and hence having short runs and many grade changes are said to loose up to 15% of their colour.

Typically, this waste colour comprises a relatively uncontaminated but much diluted version of the formulated colour. In addition to the direct costs associated with raw material losses there are further indirect costs associated with waste colour that arise from the processing and disposal of large volumes of high COD/BOD coating waste. Furthermore, the production of large volumes of dilute coating waste can also pose problems for those mills faced with constraints on water usage.

We are aware of two current approaches to coating colour recovery: membrane re-concentration[1] and recovery of pigment by selective flocculation/coagulation.[2] Both of these approaches are less than perfect solutions because in each case only some of the colour components are recovered. In the case of the membrane-based process, the soluble components are lost, whilst the selective flocculation process recovers only the mineral.

ECCI has proposed its own coating colour recovery process based on forced circulation evaporation technology.[3] Evaporation offers the following potential advantages over other coating colour recovery techniques:

- The capital cost of evaporators is low compared with membranes.
- The running costs associated with evaporators are low compared with membranes, assuming that waste heat is available at the paper mill.
- Evaporative techniques concentrate all colour ingredients.
- Evaporation techniques allow re-concentration to high solids (60+ wt.%). Typically, membrane costs become prohibitive at > 40 wt.% solids. Thus, evaporative techniques offer more options to recover and recycle coating colour to its original duty, *i.e.* coating rather than down grading to filling applications.
- Evaporators produce high quality water as a waste stream. This can be an important benefit for a mill faced with close-up issues.

In this article we describe bench and pilot-scale coating colour recovery experiments that utilize both model and real LWC and wood-free paper coating wastes.

## 2  Experimental

### 2.1  Evaporators

Bench-scale evaporation experiments were performed using an in-house built version of a conventional laboratory rotary evaporator.

Pilot-scale experiments were also performed using an in-house apparatus. The set-up consisted of a screening system, a stirred storage tank, a heat exchanger and a forced circulation design evaporator unit. In a typical experiment an aliquot of 'waste' colour was screened, then placed in the stirred tank and recirculated through the evaporator *via* the heat exchanger. The evaporator was operated at 5–10 kPa and 50–70 °C. The apparatus was run continually until the desired solids lift had been achieved. Experimental durations were governed by the volume of waste, the solids lift required and the design capacity of the evaporator (100 kg water evaporated per hour). For example, in one experiment, approximately 4700 kg of model waste colour at 30 wt.% solids was evaporated up to 2000 kg of 70 wt.% solids over a 27 h period.

It should be noted that this experimental apparatus is a prototype. In practice, such a set up would be sized according to the specific requirements of a mill. In the case of the pilot unit, energy for the heat exchanger was supplied from an oil-fired steam generator. In practice, energy would likely be recovered from the mill. The costing of such units will be very mill specific and strongly depend upon the volume and solids content of captured wastes, the required final solids and the required time-scale for the evaporation process. In general terms, experience in the use of evaporators and membranes for mineral processing applications suggests that the costs associated with such evaporative processes will compare favourably with those of membrane-based analogues.

## 2.2 Coating wastes

Model wastes were prepared simply by diluting laboratory prepared colours. Real wastes were obtained direct from a number of paper mills.

# 3 Results and Discussion

## 3.1 Recovery of LWC coating waste

*3.1.1 Model waste.* A coating colour consisting of Supraflex 80 English kaolin, styrene butadiene latex, carboxymethyl cellulose and calcium stearate was prepared at 62.9 wt.% solids. An aliquot of this colour was diluted to 30 wt.% solids, then re-concentrated to near original solids using the pilot-scale forced circulation evaporator. The rheological properties of the recovered colour are compared with those of the original in Table 1.

**Table 1**

| Colour | Solids (wt.%) | B100 viscosity (mPa s) | Ferranti-Shirley viscosity 1280 s$^{-1}$ (mPa s) | Ferranti-Shirley viscosity 12800 s$^{-1}$ (mPa s) |
|---|---|---|---|---|
| Original | 60.1 | 405 | 50 | 33 |
| Recovered | 59.9 | 455 | 57 | 36 |

The recovered colour was subsequently evaluated in blends with the original colour. The rheological properties of the colours prepared are shown in Table 2.

**Table 2**

| Colour no. | Percent reclaimed colour (wt.%) | Solids (wt.%) | B100 viscosity (mPa s) |
|---|---|---|---|
| 1 | 0 | 60.1 | 560 |
| 2 | 5 | 59.8 | 580 |
| 3 | 10 | 60.3 | 650 |
| 4 | 20 | 60.2 | 600 |
| 5 | 30 | 60.5 | 650 |

The data presented in Tables 1 and 2 indicate that the viscosities of original, recovered and blended colours were similar.

The original and recovered blended colours were pilot-coated onto a mechanical LWC basepaper using a short dwell coating head at 1400 m min$^{-1}$ and at target coat weights of 6, 8 and 10 g m$^{-2}$. Runnability observations indicated that the solids and blade pressures required for a clean blade were indistinguishable for the original and blended colours. Paper properties interpolated to 8 g m$^{-2}$ for the resultant coated and calendered sheets are shown in Table 3.

**Table 3**

| Colour no. | Percent reclaimed colour (wt.%) | Brightness base | Brightness coated | Opacity (%) | Gloss | Smoothness PPS (5 kg) |
|---|---|---|---|---|---|---|
| 1 | 0 | 69.1 | 71.1 | 89.1 | 57 | 1.40 |
|   |   | 70.4 | 72.0 |      |    |      |
| 2 | 5 | 70.2 | 71.7 | 89.3 | 57 | 1.34 |
| 3 | 10 | 68.6 | 70.9 | 88.6 | 59 | 1.48 |
| 4 | 20 | 70.6 | 71.8 | 89.3 | 61 | 1.50 |
| 5 | 30 | 69.8 | 71.4 | 89.0 | 57 | 1.44 |
|   |   | 70.5 | 72.0 |      |    |      |

These data indicate that the paper properties of sheets coated with original colour and with colours consisting of blends of original with reclaimed colour were the same.

The offset print properties for the coated sheets at 8 and 10 g m$^{-2}$ are shown in Table 4 These data indicate that the print properties for paper sheets coated with original colour and with colours consisting of blends of original and recovered colour were largely the same.

**Table 4**

| Colour no. | Percent reclaimed colour (wt.%) | Coat weight (g m$^{-2}$) | Print gloss Dry | Print gloss Litho | Print density Dry | Print density Litho | L/D | Ink setting (10 s) |
|---|---|---|---|---|---|---|---|---|
| 1 | 0 | 8.0 | 74 | 67 | 1.49 | 1.38 | 0.93 | 0.71 |
|   |   | 10.3 | 75 | 71 | 1.53 | 1.43 | 0.93 | 0.73 |
| 2 | 5 | 8.0 | 69 | 65 | 1.49 | 1.36 | 0.91 | 0.72 |
|   |   | 10.2 | 77 | 70 | 1.52 | 1.39 | 0.91 | 0.73 |
| 3 | 10 | 7.9 | 71 | 66 | 1.49 | 1.37 | 0.92 | 0.70 |
|   |   | 10.2 | 79 | 72 | 1.53 | 1.37 | 0.90 | 0.74 |
| 4 | 20 | 7.9 | 71 | 67 | 1.49 | 1.38 | 0.93 | 0.73 |
|   |   | 9.9 | 73 | 69 | 1.50 | 1.39 | 0.93 | 0.72 |
| 5 | 30 | 8.0 | 69 | 65 | 1.47 | 1.39 | 0.95 | 0.68 |
|   |   | 10.0 | 76 | 72 | 1.53 | 1.43 | 0.93 | 0.73 |
| Offset control |  |  |  |  | 1.51 | 1.24 | 0.82 |  |
| Ink setting control |  |  |  |  |  |  |  | 0.28 |

*Tack #3 Huber ink, 0–1 m s$^{-1}$.

Overall, these pilot-scale data for model LWC colours indicate that the evaporative technique for recovery of waste coating colour allows effective re-use of the spilt colour in blends with fresh colour with no significant deleterious effects on coated paper properties.

*3.1.2  Real waste.*   A sample of coating effluent (~10 wt.% solids ) from a mill running a kaolin, latex, carboxymethyl cellulose formulation was used for this work. The sample was subjected to an extensive sieving study before evaporation trials. In summary, the screening work indicated an oversize content of 0.6 wt.% of the dry material. This oversize consisted of grit and fibre in approximately equal weights. Furthermore, the oversize was effectively removed by screening at 0.38 μm.

**Table 5**

| Solids (wt.%) | B100 viscosity (mPa s) | Ferranti Shirley viscosity $1280\ s^{-1}$ (mPa s) | Ferranti Shirley viscosity $12800\ s^{-1}$ (mPa s) |
|---|---|---|---|
| 62.6 | 745 | 82 | 80 |

The screened waste was evaporated using the bench-scale evaporator to 62.6 wt.% solids. Rheological data for this recovered colour are shown in Table 5. The data shown are typical of what would be expected of virgin colour of this type and solids content. The recovered colour was blended with fresh colour (the fresh colour was formulated with SPS English kaolin, styrene butadiene latex, and carboxymethyl cellulose). The blended colours were coated onto 42 g m$^{-2}$ mechanical basepaper using a HELICOATER$^{TM}$ laboratory coater running at 400 m min$^{-1}$ with target coat weights of 6, 8 and 10 g m$^{-2}$. Paper properties interpolated to 8 g m$^{-2}$ are shown in Table 6. The data shown indicate that the paper properties of the fresh and recovered blended colours were the same.

**Table 6**

| Colour no. | Percent reclaimed colour (wt.%) | Gloss | Opacity | Brightness |
|---|---|---|---|---|
| 1 | 0 | 66 | 92.3 | 72.4 |
| 2 | 10 | 67 | 92.5 | 72.5 |
| 3 | 20 | 65 | 92.1 | 72.4 |
| 4 | 50 | 64 | 92.1 | 72.2 |

*3.1.3  Water issues.*   Several pilot-scale evaporator trials were performed using low solids waste from LWC mills. The objective of these trials was to assess the clarity and purity and hence the usefulness of the evaporated water which arises from the coating colour recovery process.

All of the samples tested were clear and with very little odour. When odour was detected it was typical of coating colours. The samples all had low conductivity, typically ~500 μS. Potentiometric titration analysis for polyacrylate dispersant indicated very low values, typically $< 3 \times 10^{-3}$ moles of carboxyl kg$^{-1}$. Typical BOD values were $< 20$ mg l$^{-1}$. Typical COD values were $<50$ mg l$^{-1}$. Spectroscopic analysis of the samples using UV and IR techniques also indicated no significant impurities.

Hence, it appears that the water recovered from the coating colour recovery process is of high quality and probably suitable for most in-mill uses.

## 3.2  Recovery of wood-free coating waste

*3.2.1 Model waste.*  A colour consisting of Carbital 90HS ground calcium carbonate, acrylic latex, acrylic thickener, poly(vinyl alcohol) and optical brightening agent was prepared at 72 wt.% solids. An aliquot of this colour was diluted to 32 wt.% solids and then re-concentrated to 70.3 wt.% solids using the pilot-scale forced circulation evaporator. The rheological properties of the recovered colour are compared with those of the original colour in Table 7.

**Table 7**

| Colour | Solids (wt.%) | B20 viscosity (mPa s) | B100 viscosity (mPa s) | Hercules viscosity (mPa s) |
|--------|--------|--------|--------|--------|
| Original | 71.2 | 6000 | 1720 | 68 |
| Recovered | 70.1 | 5700 | 1860 | 51 |

The solids contents of the two colours are too dissimilar to allow any detailed comparison. Nonetheless, the data indicate that the recovered colour had viscosity performance within the acceptable range for this kind of colour. The recovered colour was subsequently evaluated in blends with the original colour in a pilot-scale coating trial. The rheological properties of the colours prepared are shown in Table 8. The data shown indicate that there are no appreciable differences in viscosity between the colours.

**Table 8**

| Colour no. | Percent reclaimed colour (wt.%) | Solids (wt.%) | B100 viscosity (mPa s) |
|--------|--------|--------|--------|
| 1 | 0 | 71.2 | 1720 |
| 2 | 5 | 71.1 | 1680 |
| 3 | 10 | 71.1 | 2060 |
| 4 | 20 | 71.1 | 1240 |
| 5 | 30 | 70.5 | 1160 |

The original and recovered blends were coated onto 80 g m$^{-2}$ pre-coated wood-free base at 600 m min$^{-1}$ at coat weights in the range 9–13 g m$^{-2}$. The use of recovered colour blends appeared to improve runnability. The maximum solids for scratch-free running increased from 67.8 wt.% solids for fresh colour to 70.3 wt.% solids for the blend with the highest level of recovered colour.

Paper properties interpolated to 10.0 g m$^{-2}$ for the supercalendered sheets are shown in Table 9. These data indicate that the paper properties of sheets coated with original colour and with colours consisting of blends of original colour with reclaimed colour were the same. Offset print testing of these papers are shown in Table 10 and indicate that the print properties of all the samples were the same.

**Table 9**

| Colour no. | Percent reclaimed colour (wt.%) | Gloss | Brightness | Opacity |
|---|---|---|---|---|
| 1 | 0 | 63 | 90.4 | 90.4 |
| 2 | 5 | 61 | 90.6 | 89.8 |
| 3 | 10 | 65 | 90.1 | 89.9 |
| 4 | 20 | 63 | 90.1 | 89.8 |
| 5 | 30 | 61 | 90.4 | 89.3 |

**Table 10**

| Colour no. | Percent reclaimed colour (wt.%) | Print gloss | | Print density | | |
|---|---|---|---|---|---|---|
| | | Dry | Litho | Dry | Litho | L/D |
| 1 | 0 | 86 | 81 | 1.50 | 1.38 | 0.92 |
| 2 | 5 | 86 | 81 | 1.50 | 1.36 | 0.91 |
| 3 | 10 | 89 | 84 | 1.51 | 1.35 | 0.89 |
| 4 | 20 | 87 | 83 | 1.51 | 1.33 | 0.88 |
| 5 | 30 | 88 | 83 | 1.50 | 1.38 | 0.92 |
| Offset control | | | | 1.52 | 1.26 | 0.83 |

* Tack #3 Huber ink, 0–3 m s$^{-1}$.

Overall, the data for model wood-free colours indicate that the evaporative technique for recovery of waste coating colour allows effective re-use of the spilt colour in blends with fresh colour with no deleterious differences in coater operation and increased runnable solids. The resultant coated sheets have paper and print properties that are indistinguishable from controls.

*3.2.2 Real waste.* A 25 l sample of flocculated waste that had been thickened with a polymeric flocculent was used for this work. The sample had solids content 8 wt.% and a pH value of 6.9. Screening at 53 m removed ~100 cm$^{-3}$ of fibre waste from the total sample.

The sample was evaporated to approximately 23 wt.% solids using the bench evaporator. Subsequently, the sample was dispersed with a 1 wt.% active dose of sodium polyacrylate dispersant followed by liquid working with a Silverson mixer. An aliquot of this material was further evaporated to 59.8 wt.% solids. Analysis of this reclaimed colour indicated a composition of calcium carbonate (66 wt.%), kaolin (25 wt.%) and organics (9 wt.%).

The recovered waste was blended at various blend ratios with a pre-coat colour consisting of Carbital 60HS ground calcium carbonate (70 pph), SPS English kaolin (30 pph), Dow 950 latex (6 pph) and Nylgum A55 starch (6 pph). Viscosity data are shown in Table 11.

**Table 11**

| Colour no. | Percent reclaimed colour (wt.%) | Solids (wt.%) | B100 viscosity (mPa s) |
|---|---|---|---|
| 1 | 0 | 63.7 | 1600 |
| 2 | 10 | 63.4 | 1520 |
| 3 | 20 | 62.7 | 1480 |

Given the differences in solids content between the samples these data indicate that inclusion of recovered colour in the blend resulted in no significant changes in viscosity.

The colours were applied to an 80 g m$^{-2}$ uncoated wood-free base at 500 m min$^{-1}$ over the coat weight range 8–16 g m$^{-2}$ using a HELICOATER™ laboratory coater. Paper properties for the uncalendered sheets interpolated to 10 g m$^{-2}$ are shown in Table 12. The data shown indicate that the paper properties of sheets coated with original colour and with colours consisting of blends of original colour with reclaimed colour were the same.

**Table 12**

| Colour no. | Percent reclaimed colour (wt.%) | Gloss | Brightness | Opacity | Smoothness PPS 5 kg |
|---|---|---|---|---|---|
| 1 | 0 | 7 | 84.7 | 92.0 | 5.0 |
| 2 | 10 | 7 | 84.5 | 92.2 | 4.9 |
| 3 | 20 | 8 | 84.5 | 92.2 | 4.8 |

Taken together, these data indicate the opportunity to recover real wood-free wastes. More generally, these data indicate the possibility of successfully being able to recover flocculated coating wastes.

## 4 Conclusions

The preliminary bench and pilot-scale data presented above indicate that forced circulation evaporation may be a viable method for recovery and re-use of a wide range of spilt coating colours. Viscosity measurements and coater runnability observations indicate that no significant aggregation takes place during colour recovery. The recovered colours can be blended with fresh colour and used for their original duty with no significant decrease in performance. Clean water is a useful by-product of the process. Recovery of flocculated coating waste is possible.

## Acknowledgements

The co-operation of a number of paper mills in supplying coating wastes is gratefully acknowledged.

# 5 References

1    D.L. Woerner and J.L. Short, Recovery of coating losses, in *Tappi 1991 Coating Conference, Tappi Proceedings*, 251-256.
2    T. Frette, Caledonian Paper recovers china clay for re-use, in *World Paper*, 1995 (April).
3    *PCT Application WO 98/06899*, Compositions for coating sheet materials, ECC International, St Austell, UK.

# Subject Index